How to use your intuition to revolutionise risk taking

risk it

MALCOLM TULLETT

RISK IT

First published in 2021 by

Panoma Press Ltd
48 St Vincent Drive, St Albans, Herts, AL1 5SJ, UK
info@panomapress.com
www.panomapress.com

Book layout by Neil Coe.

978-1-784529-45-1

DEDICATION

To my Mum and Dad, who unbeknown to me at the time equipped me with a unique insight to the esoteric part of my career. Mum, being an old-fashioned hospital matron, had a very down to earth approach to both health and safety, from personal hygiene to using the senses that most of us were born with. Dad helped me to develop that, again not that I knew it until after he died, by making me understand my sixth sense. I have much to be grateful for to both of them and I just wish I had listened even more when they were alive and they might have been here to read this book, although I am sure they will be watching over me together and have seen every word.

To Mark

After all this time, at long last you'll get to understand what I really believe, enjoy.

CONTENTS

Chapter 1 Common sense   9

Chapter 2 A kick up the Costas   27

Chapter 3 Self-protection   45

Chapter 4 Making mistakes is normal   61

Chapter 5 Care is not just a duty   79

Chapter 6 Bad rules are not good   97

Chapter 7 Risk, opportunity and instinct   111

Chapter 8 Self and family first   133

Chapter 9 Leadership is the key   145

Chapter 10 Listen to your body and act   157

About the Author   171

CHAPTER 1

COMMON SENSE

Our 'common' sense is not common

How often have you heard the comment "but surely it is just common sense", at home, at your club, your workplace or on the TV?

Well, if it were to be the case that most things in life could be addressed using 'common' sense alone, then we would not also hear the argument that "we do not need all these unnecessary procedures that just interfere with the task at hand or stop it altogether", would we?

Accordingly, let us just examine what common sense means in relation to how we all go about our daily lives, and later I will discuss the relationship between common sense and safe work practices.

Back in 2004, I read a book called *The Wisdom of Crowds* written by James Surowiecki, an American business columnist, writing for *The New Yorker*. His opening anecdote relates to a man called Sir Francis Galton, a respected English statistician from the Victorian era. In 1906 he made a discovery when he attended a farmers' fair in Plymouth, where he became intrigued by a weight-guessing contest.

The goal was to guess the weight of an ox when it was butchered and dressed. Around 800 people entered the contest and wrote their guesses on the backs of their tickets. The person who guessed closest to the butchered weight of the ox won a prize. Once the contest had finished, Galton asked the stockman for the tickets to run them though a statistical analysis, and these were handed over as they had served their purpose and were being thrown away in any event. To his astonishment, he discovered that the average guess of all the entrants was just 1lb (0.45kg) short of the actual weight, which was remarkably close when considering it was 1,198lbs (543.4kg).

This collective guess was not only better than that of the winner of the contest but also better than each of the guesses made by cattle experts, who were precluded from entering the competition. In this instance, the common sense of the crowd led to a more accurate assessment of reality than any one individual. The fact that no one taking part knew the common sense made the process more interesting and probably why people wagered their cash to have a guess. Even those who were wildly off the mark, both higher and lower, added to the common sense by providing a counterbalance to each other, facilitating the average.

However, it must be said that Galton also suggested that to benefit from the common sense wisdom of crowds, several conditions must be in place. First, each individual member of the crowd must have their own independent source of information. Second, they must make individual decisions and not be swayed by the decisions of those around them. Third, there must be a mechanism in place that can collate their diverse opinions.

A good example of modern day common sense crowd decision making can be seen in an international television game show franchise of British origin called *Who Wants to Be a Millionaire?* All players are provided with several lifelines should they not know the answer to a particular question and one of these is called 'ask the audience'. When chosen, every member of the audience is provided with a voting button and chooses between four possible answers to the question posed.

They all make a separate and individual vote, which are then collated and displayed for the contestant to view. Sometimes, the audience fails to provide a clear answer. However, on most occasions, just one of the answers will find favour with the audience and that produces a clear result. Generally, the contestant chooses that option and in a very high percentage of occasions it is the correct answer.

The common sense wisdom of the crowd will normally win the day.

Flying in the face of the 'common' sense

To choose an answer that does not concur with the clear majority preference of the audience might also, perversely, be the right choice. This would mean flying in the face of the common sense, but everyone can be in a minority of even just one person and still be correct.

It is also worth noting that individuals cannot by definition possess this thing called common sense. In all respects, the often-used phrase "she/he has no common sense" is just a figure of speech and an oxymoron. As Galton suggested, individuals must act on their own knowledge or instinct and not be swayed by the decisions of others.

The same applies to you, your family, group of friends, work colleagues, community and even the rest of the world. Do you ever fly in the face of common sense, especially when it comes to your own safety or the safety of others? I know I do, regularly.

In fact, as part of my early naval training, my career in the Fire and Rescue Service and as a magistrate of over 30 years, I have grown accustomed to questioning the common sense of most things, as I am going to explain throughout this book. It is probably the most powerful driver to my writing it in the first place, especially as I have also often questioned the various interpretations of what lay behind the situations that have given rise to a variation of the opening statement of this book, "but surely safety is just common sense".

So, before I share with you my experiences of the world of 'elf and safety', I would like you to put yourself in the mind of that very first person who looked out to sea, saw the horizon and decided that there just had to be something beyond. Would you have wanted to fly (or sail) in the face of the common sense, with no thought for your safety or that of your crew? The common sense at that time (ie the original Flat Earth Society) was that you would simply fall off the edge. However, as soon as you returned from your voyage of discovery, the common sense immediately changed to portray the earth as round.

Ever since, the human urge to make new discoveries has relied on one person, or one group of persons, flying in the face of the prevailing common sense at that time. In many cases, especially in those early pioneering days, safety was not a high priority, but as civilisation has progressed, we all learned from our disastrous first attempts at exploration and developed better and better methods of protecting ourselves.

One such person was Joseph Strauss, the chief engineer on the Golden Gate Bridge construction project which started in 1933. Flying in the face of the common sense at the time, he made safety his highest priority on this treacherous project. He made everyone working on the project wear hard hats and fall arrest lines. He also spent a then unheard of $130,000 on an innovative safety net that was suspended underneath the bridge. Although the 'bridge men' were considered fearless daredevils and used to working without any safety precautions, Joseph would dismiss anyone who refused to wear it on the spot.

The industry norm back in the 1930s was that one man (women were not employed in mainstream construction activity) would lose his life for every $1m spent. The bridge cost $35m to build and accordingly the common sense expectation was that 35 men would lose their lives. This was unacceptable to Joseph.

He insisted on all the latest safety innovations, including the safety net, to reduce the potential loss of life. Nineteen men fell into the net accidentally, yet none of them died, and this small group of lucky survivors became known as the Halfway to Hell Club. All his safety precautions, including the safety net, fall arrest lines and hard hats, were considered the most rigorous of the time, and he enforced them rigorously too. Other innovations included a sand-blasting respirator mask, glare-free goggles that prevented 'snow blindness' caused by the sun's reflection off the water, and special hand and face cream that protected the skin against the constant high winds. He went as far as to establish a field hospital and provide carefully formulated diets that helped to fight dizziness.

Unfortunately, 11 men did still lose their lives but 10 of those were killed in a scaffold collapse. Notwithstanding the unnecessary loss of life, the project was still the safest at that time. The anticipated fatality rate (even that sounds awful) was cut by two-thirds and 24 men, at least, owe their lives to an individual who was not prepared to go along with the prevailing common sense.

The educated idiot

However, that does not mean that everyone who flies in the face of common sense is correct; far from it, and generally the perceived wisdom is correct, up to the point where it is disproved. You must always remain cautious when going against perceived wisdom. It is not just about knowledge though, as sometimes we come across the 'educated idiot'. I recall an incident on a coach while travelling to a rugby match whereby a very intellectual member of the team was having trouble opening a window, which required him to simply squeeze two parts together so that it would slide open. He could not even work that out. It was not that he had no common sense, although he patently did not, he was simply an educated idiot.

You must have come across such a person. We have all come across the situation where the organisation for which you work needs to fill a supervisory position. Instead of promoting from within the ranks, the boss brings in someone who is 'qualified'. They might have a first-class honours degree in some sort of 'ology' but have little or no actual experience. I have seen this many times and generally the common sense will prevail to get things done despite, rather than because of, the presence of the 'idiot'.

Actually, I feel quite sorry for anyone 'dumped' into this position, as all too often they are pressured to perform and seem unwilling to seek advice from long-term members of the team for fear of appearing as a failure. It is, of course, possible that the team members resent the imposition of the new supervisor, who knows little about the task. Some of them might even be only too glad to see him or her fail and be shown the door.

Conversely, I have also seen promotion from within the ranks of those who may have had the technical skills but lacked the people or organisational skills to lead the team. One variant of this is known as The Peter Principle, which states that "in a hierarchy every employee tends to rise to [their] level of incompetence" and in time "every post tends to be occupied by an employee who is incompetent to carry out [their] duties".

Either way, morale drops and either production suffers or, worse, a poor safety culture develops. My corollary is that all persons taking on a new leadership role, at whatever level, should receive adequate and appropriate training for that role before being expected to take on the role. This is what happens in the military and blue light services, so why should it be different in industry and commerce?

You might also think it would be a good idea to train parents before they have children. I couldn't possibly comment other than to say the textbook would probably be the longest one I've ever seen.

Cotton wool

Training, or to some degree, more importantly, education, is a funny thing and an emotive subject. For instance, in my opinion, the education system, particularly in the United Kingdom since around the 1960s, has tended to wrap children in cotton wool and overprotect them from any form of harm.

Although understandable to some degree, in an ever more dangerous society, I believe that this has the countereffect of reducing a child's ability to protect himself or herself. This includes those risks that we all face daily, as well as those we only come across occasionally. It seems like the ultimate consequence(s) of every risk known to humankind is now what drives our actions, in the so-called western world at least. You will have worked out by now that I feel that the likelihood of that event occurring at all has been missed from the considerations. Neither do they take account of any forms of control, such as our own ability to take care of ourselves.

To illustrate the point, while I was a trustee of a professional body for safety practitioners (the Institution of Occupational Safety and Health – IOSH), a head teacher in the West Country banned all of the children at the school from playing 'conkers' (in the UK, horse chestnuts are also known, colloquially, as conkers) and the press got hold of the story, under the headline of 'conkers bonkers'. This sparked me and my fellow trustees into action.

We instructed the chief executive to write to the head teacher, explaining that the long-established practice of conkers (ie threading conkers on to strings and taking it in turns to try and break your opponent's conker) was not worthy of prohibition. It was hardly likely to lead to a child's death or major injury. Accordingly, the letter was duly sent, along with a suggested risk assessment that indicated the risk to be 'trivial' with the likely outcome being hurt pride if one of the participants failed to achieve success in the competition.

To our utter astonishment, back came the response with a counter assessment suggesting that the likelihood of failure while attempting a violent action of destruction was extremely high. It went on to suggest that with most conkers being broken after the second or third attempt, the potential for knuckles to be hit was significant. Even worse, apparently, pieces of broken conker could fly off and potentially hit a child in the eye. So, instead of banning the 'sport' altogether and due, in no small part I assume, to the adverse publicity received, the head teacher provided goggles and gloves for those children that still wanted to play.

It was just getting worse and my fellow trustees and I became incensed. It was our unanimous view that his response to what had become, previously, an everyday (during the autumn) occurrence in the playground was completely over the top.

We decided to stage our own conkers championship, in front of the media. The sight of adults wearing full PPE, including industrial goggles and boxing gloves, for effect only, was hilarious. It too featured in the headlines. We all believed that this stage-managed spectacle would have discouraged that head teacher, at least.

Alas, to this day, stories of head teachers continue to appear in the media. Some have even gone as far as to suggest banning this 'dangerous sport'. The common sense continues to refer to this as 'conkers bonkers' and it continues to fill me with sadness.

It is yet one more reason why I have decided to write this book.

Dodgems

Luckily though, I also believe that this is a truly English phenomenon. I do not think the other nations of the UK are quite as overprotective and it is probably not a problem in the north of England either. They constantly refer to us as 'soft' Southerners and it would seem they have a point. I wonder how other countries would see this wrapping of our children in cotton wool. What do you think?

While living and running a safety practice in Spain, I had the absolute pleasure to see how they bring up their children. I know what you are thinking, how come I was running a safety business in Spain, but more about that in the next chapter. What I loved about the Spanish way of life was their love of children. Yes, that included allowing them to experience the fun that is associated with an element of danger. Just like my experience of South African families, the adults did their thing while the children went off to play.

Once upon a time, when I was at junior school, our parents had the same ethos and I remember going to the funfair (an outdoor event featuring games, rides, exhibitions etc). One of those rides was the bumper cars, where we, quite literally, drove our own electric cars and bumped into each other. It was great fun and the kids just kept going back for more. Somehow, over time and because of the actions of the cotton wool brigade, this all changed.

In the UK, at least, they are no longer even called bumper cars. They are now called dodgems and any form of bumping is prohibited. In fact, all cars are expected to travel in the same direction to avoid the likelihood of head-on collisions. Even worse, the cars can only be driven by adults and the kids must wear a seat belt. I just fail to see how the children of today can have as much fun as I did.

Not so in Spain, or at least not when I lived there. Adults were not even allowed in the car, as they were considered a children's ride. The cars were provided with a seatbelt, although it only needed to be worn if the child felt more comfortable and, of course, safe. The children could drive their car freely. The only restriction was the attendant keeping a lookout for older children being reckless. Even then, the response was simple. The ride would be stopped and the offending child spoken to, in front of all the others. It was very effective and all children were then left to carry on having fun.

"Heresy," I hear from some of my colleagues. However, many of these people have only learned safety from a book or at college. No, I am not suggesting they are all educated idiots. Far from it, as they are obviously concerned about child safety, as am I. However, I do not understand those who do not accept that, as a parent, we must ensure that our children are capable of understanding, experiencing and dealing with risk.

There must be an acceptance, surely, that we must all be able to deal with daily risks, including our children, if we are to live with it appropriately. Most of us are equipped with the natural ability to assess risk, in our head, in a split second. We hone that skill as we age and hopefully acquire some wisdom to be able to interpret our assessment correctly. In short, even at a very young age, most children can use their senses to see, hear, taste and feel danger, with smell being the strongest of all senses. I also believe that a sixth sense exists, which I will call our 'tuning fork' and this will be discussed in greater depth in chapter 3.

The assessment process is very simple and requires just two inputs. First, you need to have an idea of how likely an activity or event is to occur. Then, you will need to consider what is likely to happen when that activity or event does occur. As you will see, there is nothing definite in this most basic form of assessment. It is a judgment call. It is why virtually every human being on the planet, except for those who sadly lack capacity, can and do undertake this process, continuously, while they are awake.

Be 'reasonable'

During my career in the Fire and Rescue Service, I had the good fortune to be selected for what was called the 'long course'. It was so titled as it entailed 13 weeks of specialist fire protection (SFP) training. Very early in the course programme was an entire day devoted to the discussion of the word reasonable and the phrase 'reasonably practicable'. At the time, we all thought that a whole day was excessive. However, it has stood me in good stead ever since and is an often-used word that crops up daily in my professional life.

The legal interpretation of the word reasonable was introduced into English law during the Victorian era. It is still an important concept of British law, being later incorporated into both case law and legislation. The meaning of 'reasonably practicable' is best described by reference to a set of a scales. On one side of the scales is the risk and on the other side are the steps needed to counterbalance the risk. So, in other words, it is a balance between the risk or potential for harm on one side and the cost of preventing or mitigating that harm on the other. Obviously, if the risk is great, then cost, in its wider sense of money, time and effort, needs to be greater to reduce it or even remove it completely.

However, this cost-benefit analysis is not always simple. Although it is not necessary to remove all risks, even if a low level of harm can be reduced using easier and more proportionate steps, then they would be considered 'reasonably practicable'. First then, the level of risk needs to be assessed. After that, you will need to consider what can be done and then what is 'reasonable' to do in the circumstances.

So, who decides if what you have done was 'reasonable'? Well, you will be pleased to know, I am sure, that it will be a 'reasonable person' and the 'man on the Clapham omnibus' is that hypothetical ordinary person. The character of this omnipresent and mythical individual is a reasonably educated, intelligent but nondescript person against whom your conduct can be measured and who is used to set 'reasonable' standards. The person has no specific gender other than in legal text which uses the male gender to include both the male and female genders. The Clapham omnibus is simply an old-fashioned term for a bus.

The reason why that analogy was used to explain 'reasonableness' was, in my opinion, to identify someone with the common sense of the people and a good view of what was going on around them. It remains one of the most important of all concepts, as it suggests that the determination of what is 'reasonable' should always be based on the common sense of the people.

It is the total opposite of those with extreme opinions, as well as those having no real-world understanding or experience of life. It does not include those who are incapable of thought or academics or specialists with a narrow field of expertise, although their sense of the world might be useful in research and highly technical issues.

What is not 'reasonable'

If you like a little flutter on the horses or similar, you might disagree with me. However, I do hope not, as I would suggest that it is perfectly reasonable to gamble occasionally, though not to excess.

It must also be wrong to gamble with your or someone else's life. We all make instant decisions, every single day of our lives. As mentioned before, these judgment calls need to be balanced and not in themselves just a gamble.

To explain, I had to attend an incident at a west London train siding one early Sunday morning. An unfortunate individual had been crushed to death between the floors of a railway car transporter. Although there was nothing that could be done by us other than to extricate the body, the history of the event is worth exploring.

There should have been two people loading and unloading the cars that morning. The normal and safe operating procedure required both to be in attendance. One would raise and lower the ramps and the other would drive the cars on to the platforms. The fact that this operation was being undertaken outside of normal operating hours was also an important factor. It meant that they were receiving an enhanced payment for weekend working.

Their assessment was based on the balance between risk and cost. Unfortunately, they overplayed their hand and prioritised the reward. They got greedy. They agreed between themselves that they could do this alone. Each would take it in turn to have a Saturday night at home or with friends. They both continued to receive their enhanced wage packets for some time, potentially years, without incident.

On this fateful morning, however, the lone operator slipped up. Literally. He had been operating the ramp and driving the vehicles all night, so was probably also tired. He slipped and fell between the upper and lower ramp and was unable to extricate himself. No one knows, obviously, but I do sincerely hope that he knocked himself out. Had he been conscious, he must have screamed out in both fear and excruciating pain as that upper ramp squashed the life out of him.

The problem was that he was working alone and no one else was around to hear those screams. His mate was probably tucked up nicely in bed. All it needed was for someone, anyone, to press the stop button. These guys took a gamble which was very rewarding for some time but the inevitable happened. One of them paid for it with his life.

Any 'reasonable person' must agree that the assessment they did was not 'reasonable', let alone lawful, and would have said, "It's just not worth the risk."

When even 'reasonable' does not apply

However, there are some instances where being 'reasonable' is not enough. For instance, when it comes to issues like the guarding of machinery and environmental protection, the risk of doing just what is 'reasonable' is not enough. How would you justify assessing that no guards are required for a bench saw? How would you argue for the filters to be removed from toxic emissions? How would you argue for untreated effluent to be allowed to enter a stream?

As I am sure you will understand, in the case of a machine, unless it is totally impracticable to use it with a guard fitted, there must always be a guard in place. When it comes to environmental protection, if the environment and wildlife can be protected, it should be, at 'virtually' any cost.

One of my clients produces architects' models which require very intricate shaping of miniature buildings using balsa wood, and, accordingly, the wood needs to be cut to size in a very specific way with a small bench saw. I explained that the saw had to be provided with a guard to comply with the statutory requirements and was told that it was impossible to use the equipment with any form of guard in place. Under normal circumstances, the functional standard would be 'reasonably practicable' but in the case of machinery, the law (in the UK) specifies a much stricter standard of simply 'practicable'. So, as it was totally impracticable to use the machine without a guard, it fell into that grey area where although the machine had to be used without a guard, the work still had to be done safely.

In chapter 7, I will talk about the hierarchy of risk controls. However, just for now, I will keep it simple and explain the practically of considering the other controls. It was agreed that the machine would be moved into its own dedicated space, clear of other work. While the saw was not being used, it would also be provided with a guard that was padlocked. This was to ensure that no one could run their hand over the top of the saw blade, even when the blade was not spinning.

Only a small number of 'competent' members of staff would be authorised to access to the key, and we will examine what is meant by the term 'competent' later. It was also impossible to use gloves, even fingerless ones, as the operation required fine dexterity and any form of hand protection would have hampered the work. It was further agreed to provide miniature push sticks with pins to secure the material being cut. Regular refresher training of the authorised users of the equipment was also provided in order to maintain their competence.

An even more rigorous approach is required for environmental risks, which go a long way beyond the common sense of the crowd. I mentioned earlier that if the environment and wildlife can be protected, it should be, at 'virtually' any cost. This very high standard, known as BAT, requires the adoption of the 'best available techniques', although even then, the caveat is that they should not entail excessive costs. Simply put, this means that the highest standard of protection is utilised and if it has already been done somewhere else in the world, then it must be done in all like circumstances. The protection devices or systems must be provided to the same or a better standard. Furthermore, although the word excessive is both subjective and arbitrary, legal precedents have been made by the courts around the world.

Custom and practice

One area of concern when applying common sense is the age-old phrase "but we've always done it that way and never had any problems", which sounds on the face of it quite compelling. However, my initial reaction is that it indicates a negative safety culture.

Just consider for one moment a deck of playing cards with a single joker, which I would like you to think of as something going wrong, like a personal injury accident. Every time a card is turned over it represents one working week, so there is a whole year's worth of work in a single pack. The approach being taken by those making the excuses for not doing something safely is that nothing has gone wrong yet. The operative word in the sentence is *'yet'*, which is where the joker comes into play, and the gamble is that it will not be the very next card turned over. Every year, a new set of cards is used, and again, the joker sits at the bottom of the deck, hopefully with the other joker that has not *'yet'* been turned over. The longer the game continues, the more chance there is of unearthing the joker, unless it is removed by positive safety practices.

Many years ago, in the UK at least, a particular custom and practice known as 'grandfather rights' was outlawed. The practice allowed those with long-standing experience in an industry to train others. This arose from the unhealthy adherence to the principles of the previous analogy. The 'old hands' were believed, wrongly, to have got through their service without adverse incident, so they must know what they were doing and getting it right. What the supervisors and managers did not know was that, again, the joker had not *'yet'* been played. It had been avoided, more by luck than judgment.

These so-called 'grandfathers' may well have been performing their task(s) so unsafely that they were covering up repeated failures, otherwise known individually as a near miss, or more accurately, a near hit. They had probably been doing it for so long that it was second nature. Their unsafe practices would then be passed to the rest of the team and, significantly, to any new starters including young apprentices. Alongside the skills, knowledge and experience, they would also pass on the bad habits but the novice(s) would be unaware of the dangers involved and unable to either avoid or mitigate them without the time-served experience. Being put 'under the wing' of the 'old hand' perpetuated bad and potentially life-changing or life-threatening habits. Not the best start for a young person just starting their working life.

So, common sense needs to be treated with caution. It can be a force for good and can lead to both accepted bad practice and a general acceptance that all will be fine, simply as nothing had gone wrong 'yet'. Remember, it is your choice. You can either fly in the face of common sense to achieve better and hopefully safer outcomes or you can fly in the face of common sense to perpetuate unsafe behaviour.

Let us return to the incident of the car transporter. Why was that dangerous 'custom and practice' never stopped? Was a supervisor or manager never on site to see what was happening or, heaven forbid, were they in on the fraud? Did anyone else know what the two guys were up to and, if so, why did they turn a blind eye on the practice? What would you have done?

Let us both examine these issues and other difficult questions throughout the rest of the book. The villain of the piece will not be revealed at the end, like a novel, as it could be you, or someone you know, in plain sight. We will examine the evidence, establish the modus operandi and motive together to find your truth.

CHAPTER 2

A KICK UP THE COSTAS

The cultural side of common sense

Back in the mid-1990s, I had the good fortune (in some senses of the word) to move out to Spain, where I purchased a small estate agency with my first wife and her brother and sister-in-law, along with their daughters and my stepson. We had all been on holiday there during the preceding few years and, like many others before us, just loved the lifestyle. We were about to find out, however, to our cost, that living the dream is not always the same as working the dream. Hence the chapter title.

At that time, which was shortly after I retired from the London Fire Brigade, my wife had a relatively successful art gallery and picture-framing business in a small village in Surrey and I had just started a private fire and safety consultancy. The timing was good and both businesses landed on their feet. However, we decided to sell the house and move to Spain. We were to learn, very quickly, that the reality of living and working in a foreign land was to prove a massive culture shock. Although the lifestyle was great, it had some very mixed blessings.

In terms of the work-life balance, our first lesson was to find out what the word *mañana* really meant. While on holiday, there were plenty of *mañana* events but we were on holiday, so one day simply blended in with the next anyway. We were now living the dream though, so we soon found out that it just meant not today, or worse, some fictional time in the future. It was not how we were used to doing things and the fact that the Spanish generally work to live did not feature in our plans.

The Protestant work ethic (in sociological terms) of the northern Europeans, including us Brits, supports the value attached to hard work, thrift and efficiency. As we were all about to find out, the Catholic work ethic might be better termed the Catholic leisure ethic. It was immediately apparent, though we just never saw it while on holiday, that the Spanish people and the Catholic Church took the view that having 'enough' material wealth allowed for a greater development of the person. Probably, from ancient times, the likes of Socrates and Plato would have pondered the natural world and looked favourably on what must have seemed to them the innocent and empowering pursuit of leisure and the arts. It is a pursuit that those who live in and around the Mediterranean pursue to this day.

So, the common sense of the population was already working in a different direction. The saying "when in Rome" (in fact, we were in Málaga province, on the Costa del Sol, in the region of Andalucía) meant that we had to "do as the Romans (well, Andalusians anyway) did". The problem was further confused by the integration of the Moorish (North African) culture, which undoubtedly played a fundamental role in the creation of the multicultural profile of this unique region of Spain.

Although Andalucía was once occupied by the Romans, the Moors dominated much of the Iberian Peninsula for several hundred years, instilling their culture in everything from the science and architecture to the food and drink. Once the Catholic monarchs of Spain reclaimed Andalucía, a distinct blending of cultures and unique customs ensued, which remains to this day, such as *tapas*, flamenco, late nights and street parties, known as *botellónes*. I cannot see how, in this modern world, bullfighting continues to attract the crowds. It just goes to show how the Andalusians react to being told what to do by outsiders, like the European Commission.

The Andalusian way of life is distinctly laid-back and drawn out, just how the locals like it and just as I did initially.

Fun and safety

So, do you recall in the last chapter that I spoke about those dodgems? The Spanish love to have fun and never let this little thing called safety get in the way. Just ask yourself a very simple question. When you were young and playing with your mates, where did you have the most fun? It was probably when you were out of sight of your parents. Now ask yourself another simple question. What sort of things did you get up to, especially if you are a boy, not that it ever stopped the girls? Well, that probably had something to do with how far away from your parents you were and the further away, the more adventurous. Or, as your parent or guardian might have said, 'naughty', as it is most likely that your adventures had an element of danger attached to them.

This is the part where I loved the Spanish people. The children were brought up as the centrepiece of family life, and traditionally they ate at the table with mum and dad and the rest of the family.

They were encouraged to get involved in the more robust games and pastimes, like going to the fiesta. These fiestas were normally religious festivals and most had funfair activities, like our traditional bumper cars (dodgems in the UK). Most of these activities and rides had an element of danger attached to them, which made them so much more fun and the children loved them. I never saw one child wearing a seat belt in the bumper car and they were doing all the driving. No tears and no tantrums.

Spanish family life was different, that is for sure. The idea was that children should learn that life was a risk. They knew instinctively that if all risk were to be avoided there would be little fun and few opportunities. This is exactly why the 'conkers bonkers' mentality simply wraps children in cotton wool.

Not so funny (well, a little bit funny)

There were several occasions when I witnessed the reverse mentality, side by side with the laid-back culture of the Andalusians. On one occasion, early on in my adventure, I was walking along the *paseo*, a seafront footpath, where some major drainage work was taking place in the roadway. It lasted for many months and I was absolutely astounded to watch what was effectively an enormous U-bend being inserted into the two ends of the pipes, where they came together from each end of the road. This was due to their levels not being correctly plotted, by the engineers before the work started. Still, although I suppose it must have worked out in the end, I would have thought that getting the levels right first would have been a normal civil engineering requirement.

On a separate occasion shortly after, while watching those same works, I had this funny feeling, just like a tuning fork going off inside me. I just knew something was going to happen and probably something unsafe. I stopped, looked around and saw something unfold in the distance. Anyone and probably everyone else on the *paseo* would have seen, let alone the construction workers had they looked. To my horror I saw a mechanical digger working immediately adjacent to a concrete barrier. The barrier was being used to separate the work area from the live roadway and was about 1m high. Unfortunately, the digger was so close to the barrier that the elbow kept crossing over into the space above the roadway every time the digger needed to shed its load into a waiting lorry.

A bit further down the road I saw a bus. I could instantly see what was going to happen unless there was an operative standing by to stop it, which I hoped there was. Nope, no such luck, or should I say foresight. It was as if the now inevitable comedy of horrors being played out in front of my eyes had been choreographed. I was far too far away to stop anything, so I just carried on watching. Incredulous. The bus driver must surely have seen what was to come. Nope, he was oblivious too. At the very moment that the bus was passing, the elbow of the digger swung across in front of the oncoming bus. I feel sure that they could not have choreographed it better had they tried. The digger elbow went straight through the bus window.

Both drivers were totally stunned but luckily unharmed. Their pride was obviously hurt. It just kept getting better though. As I got closer, I could see that each driver was blaming the other, with what seemed to me to be a very Spanish exchange of expletives. At that time, I could not speak Spanish but could discern, from their body language and tone of voice, that neither of them was best pleased. Everyone involved then just cleared up the mess and they went about their business again. The local boys in blue (police) arrived and were totally disinterested with the damage. They simply unblocked the traffic mayhem and they too went about their business. Just one more day in paradise.

The story of things to come

From my early days in Spain, I witnessed some very strange happenings. I still wonder to this day why I kept going in this *laissez-faire* environment. I must have simply put it down to the call of the carefree lifestyle but even that was fraught with some very strange customs and practices.

For instance, for some reason we combined two very stressful activities into one memorable event. We took possession of our first house and exchanged contracts on the estate agency at the very same time. It seemed like a good idea. We were introduced to the *abogado*, a legal professional who is in charge of the direction of the parties in all types of public and private legal proceedings. We were introduced to him by our new bank manager, so what could possibly go wrong? We trusted in the assumption that these professionals were acting in our best interests. You know that tuning fork I mentioned earlier? Well, it started to vibrate again and I really should have listened to it. However, not knowing the language at that time, or the practices, I put my faith in the professionals and ignored it. Bad move.

As with most conveyancing, it is customary to make payments for certain services up front and Spain was no different. So far so good. The vendors needed to be paid, as did the *abogado*, and the utility companies had to be contracted. Again, so far so good. The problem came when, armed with my new cheque book, I asked who the first cheque should be made out to and was politely informed El Portador. Now, for the uninitiated or Spanish speakers, I hope you are with me here. I had always thought that sort of comment related to someone's name, a bit like El Cid. Just for a bit of background, this is what the Moors called Rodrigo Díaz de Vivar, a Castilian knight and warlord in medieval Spain.

Anyway, back to my tale. I duly signed my first cheque in favour of this person that I believed was called Mr Portador. To my surprise, the second cheque was also to be made out in favour of Mr Portador and likewise the third. At this point, my tuning fork was vibrating so hard to the point where I was on the verge of becoming nauseous. I just had to ask. So, "Who is this Mr Portador?" I said sheepishly. "No," said the *abogado*, "it is no Mr anything, it is Spanish for the bearer" otherwise known as 'cash'.

It was here that I had my second Spanish lesson and another one that I will never forget. I also understood a bit of Latin that day, too. Caveat Emptor means 'let the buyer beware', which means that the buyer (me) assumes the risk that a product or service (legal advice) may fail to meet expectations (advance warning) or have defects (maybe the *abogado* was the illusive Mr Portador). I still wonder.

This was not the end of my rough handling by the Spanish though. After about a year of hard knocks and learning from my mistakes, I was approached by a Spanish acquaintance with a business opportunity. It was directly related to my own safety business and obviously I was all ears. Oh yes, I forgot to mention that I was also commuting back and forth to Old Blighty (England), once a month. My UK safety business was in full swing and being handled on a day-to-day basis by an ex-fire service colleague, who had also become a trusted employee. The opportunity was explained, it sounded interesting, so I enquired further. Another bad move.

At that time (mid 1990s), there was no professional safety expertise in Spain, as the national, regional and local governments all treated it as either an insurable risk or nothing to do with them. In addition, following a serious incident their role would simply be to intervene and prosecute the offender(s). However, if you recall, Spain had entered the European Union (EU) around the mid-1990s, coincidentally the same time as I became a Spanish resident. As a quid pro quo for the massive infrastructure investments made by the EU, Spain had to raise its game in the safety arena, amongst others.

Until that time, the only accident prevention organisations were a type of Friendly Society, known as *Mutuas*. These not-for-profit organisations were set up to administer compensation for workplace injuries, on behalf of the Spanish government. They were not equipped, at that time, to interpret the EU regulations and directives related to workplace health and safety. This is where my Spanish acquaintance came forward. He was associated with one of these *Mutuas* called Solimat and suggested that I speak with one of their senior managers.

The first meeting was one of several lunches, with customary brandy and cigars. Initially, it was to see if we could become friends, which was the Spanish way. Seemed like a fantastic way to do business to be honest. The next meeting was to see if our wives could be friends, which I found a little strange as that is not the way I was used to doing business in northern Europe. After that, we met again, without the wives this time, to see if there was an understanding of each other's positions in general terms. After a couple of hours, it was obviously too late in the day to agree heads of terms, so we had yet another meeting for that purpose and only then could we move on to our final meeting. At long last, we sat down to another lunch, with more brandy and cigars, to agree a joint venture.

Now the fun and games really begin

I was about to learn my third lesson about the Andalusian (Moorish, remember) culture regarding Caveat Emptor. It turned out that in the case of my Anglo-Spanish venture, formally incorporated as IRM (España) S.L., the English partner (me) put all the money in and the Spanish partner (a German, representing the *Mutua*) took all the money out. Actually, that is not strictly true, as he was aided and abetted by the system.

In Spain, in the area covered by the Junta de Andalucía at least, a system of formal and very bureaucratic accreditation existed. Firstly, unlike the UK where professional qualifications are accepted generally once they are agreed by a national body, in the regional governments of Spain, these qualifications needed to be individually assessed in the relevant colleges. Accordingly, as no such colleges were in existence in Spain at that time, I had to seek what was known as a Postulate of the Hague Convention, or proof that my credentials were "true, as the basis for reasoning, discussion, or belief".

I needed this before I could even sign the application papers for permission to start trading as a bona fide health and safety practitioner, and yes, it took several months. Remember, the Andalusians are renowned for their laid-back – it will happen when it happens – approach to life and that includes work. As soon as it came through, I made the first application and noted, to my horror, that the state had already determined the structure of a health and safety practice. Unlike the rest of the EU, Spain had unilaterally decided that a medical facility was part of the mix and I needed enough space to set this up, for occupational health examinations. In the UK, occupational health practitioners have little to do, directly, with health and safety consultants.

This meant that I needed to find a site large enough to house a medical practice, training facilities and offices. That was not all though. I had now been considered acceptable in terms of my competence to establish the company. However, I was not authorised to practise, as only those with a Spanish practice certificate could do that. I was much better qualified than most in the country at that time but I had no choice but to employ one of the few graduates from the very first cohort. The so-called 'expert' I chose was, I felt, the best of the bunch but had no practical experience whatsoever. So, in addition to the cost of the premises, including refurbishment, the salary of my Spanish co-director and a rooky 'expert', I also needed the services of a full-time occupational health nurse and the retained services of a doctor.

As I am sure you can appreciate, this was all bleeding my UK practice dry, but somehow I saw it all through to the interim approval and, at last, all would now be good to go. However, Spain is well known for its bureaucracy and my situation was to be no different. I had my credentials checked. I set up the business and the premises as required. I employed the so-called 'experts', who had been busy creating the necessary client documentation and several training courses. All had been checked by the Junta (regional government), so what could possibly go wrong now? Well, that interim approval was nothing at all to do with permission to trade. Can you believe it?

We had already been introduced to our new clients by the *Mutua* and I was expecting to hit the ground running. So, I asked where I would get this permission from. I could not believe my own ears when I was told by the inspector that issued the interim approval that it was him. In hindsight, another stupid question left my mouth. "Pray tell," I asked, "how do I go about getting that permission?" Without a pause or stutter, he said he had to carry out an inspection. "Why?" I asked, "could you not have done both at the same time?" He said because it is a different department and a different form, plus of course another fee. I bit my lip and requested another inspection and agreed to pay the fee. "But," he said, "you haven't filled out the form." I duly completed the form, which was the same as the previous form but with a different department and reference number and posted it (no electronic submissions in those days).

I had to wait two more weeks before receiving a response and a further two weeks before the inspection, which was no different from the first inspection and he even had the same questions. I asked when I would receive permission to trade and heard something like, *"el día después de mañana."* I was getting used to this phrase by then and just knew, instinctively, that it did not actually mean the day after tomorrow but rather at some stage in the future. I really could not wait any longer. So, together with my Spanish-German co-director, we decided that as we already had clients waiting, we would get started. We also realised that any invoicing would need to wait until full approval came through.

The problem was that I was ploughing money in, continuously, with no pay back and the International Guarantee was getting very close to being exhausted. I had to decide, soon, if my English company was going to go down to keep the Spanish one going. Deadline after deadline came and went. In the end I had no choice but to pull the plug.

It wasn't just me

While my nightmare was playing out, one of the bars next to the estate agency was also about to find out about the rough end of doing business in Spain. The bar was blessed with some orange trees at the front of its terrace, which contributed to attracting clientele. They were, however, starting to get overgrown, encroaching into the bar and damaging the awning. The owner, who had now become a friend, wanted to get them trimmed but he could not touch them as they were the property of the *ayuntamiento* (town hall). He went along and waited in the queue for a couple of hours to be seen, before explaining to the clerk what he wanted to do. Apparently, the clerk was unsympathetic. He explained that my friend and neighbour would need to wait until the trees were due for trimming but could not tell him when that might be as it was a different department. You will have heard that one before.

Anyway, every few weeks the bar owner returned to make the same request. Every time his 'complaint' was physically placed at the bottom of the filing tray. His temper was now getting frayed and on one occasion he lost it. His temper that is. He raised his voice just a little too much and demanded some action. He was arrested at gunpoint and escorted out of the building. That was not quite the action that he had expected. On his release from the police station and very dejected, he returned to his bar. He found that the municipal police, employed by the town hall, had beaten him to it and padlocked it shut. On the door was a notice explaining that the *supermercado* (supermarket) had been shut down by the *ayuntamiento* as the owner had failed to obtain permission to trade as a *supermercado*.

Now, although this was a complete surprise to my friend, having been trading as a bar for several years, he found himself on the wrong side of bureaucracy. It took him nearly six months to get it resolved and only after he managed to get an audience with the mayor. He agreed to buy a new set of awnings from her brother. Of course, that had nothing to do with the reinstatement of the bar as a bar.

Shortly after the bar reopened, the *ayuntamiento* came along to trim back the orange trees, as originally requested. This operation was undertaken using a chainsaw under the lowest branch, leaving just the trunks. It would appear, therefore, that the common sense in Andalucía was to just pay your taxes (or cough up for a new awning from the mayor's brother) and go with the flow. If that meant you had to pay for your supermarket licence when you ran a bar, then so be it. Risk management in Spain is a funny old business.

What about the law?

Spain has plenty of laws (Real Decrees) and, as you will have seen, plenty of bureaucracy to go with them. Unlike the UK, most of the Mediterranean countries maintained a codified system of law, which had a rule for everything. In the UK, one could do whatever they wanted if there were no law(s) that said one could not.

Step in the *gestor*, a professional person, though not required to be a licensed attorney (*abogado*), who deals with administrative bureaucracy on behalf of a client. They are listed in the Páginas Amarillas (Yellow Pages) as *gestorias*. It seems strange to have the need for such a person but I am sure that my bar-owning friend could have found use of their services had he have known about them at the time.

These *gestorias* were also equipped with all the workplace health and safety decrees and administered them in a very prescriptive way, without any knowledge of the underlying principles involved. For this reason, their clients were, in most instances, equipped with a paper mountain of must and must not items, otherwise known as a tick sheet. For those very few clients for whom we did manage to provide services, it was very clear that the tome of papers would have been better used being torn into shreds and used as toilet paper.

In each case, I had to explain to my own 'experts' that the written word had never stopped an unsafe act or condition. On the contrary, too much of it can cause the problem. Less is usually more. The key, however, is implementation and in most instances the workers did not even know this 'toilet paper' even existed. Even the workplace inspectors only ever looked to see that the organisation being inspected had the necessary paperwork. That is one of the biggest problems with a bureaucratic system.

The 'law' was also being enforced in a very officious way by the enforcers. I am told that it has changed somewhat since I was there, although the road traffic laws were always enforced correctly, at Christmas anyway. The local police realised that enforcement had money-earning potential, as all fines were the property of the authority and could be divided up to pay a Christmas bonus. During the rest of the year, motorcyclists, for instance, were required to wear a crash helmet but the law did not state where they were to be worn. Can you believe it? The young riders took advantage of this and strapped them to their elbows. Very effective.

Strangely enough, income generation from criminal enforcement, including breaches of health and safety legislation, seems to be the order of the day now in many countries. In England, the concept of a 'fee for intervention (FFI)' was introduced in October 2012. This required duty holders, who were considered by the inspector(s) to be guilty of a material breach of health and safety law, to pay for the time of the inspector(s). This includes all work to identify the breach to help offending companies to correct their failings, any investigation and even the time it takes to take more serious enforcement action, such as prosecution.

From the outset, as a consultee to the then proposed new regulations, I argued against this money-making regime. It seemed to me that the inspector(s), who normally work in pairs, would be pressured into finding 'material' breaches that would not normally lead to other forms of enforcement, beyond an advisory letter. It soon became apparent that the number of 'formal' Improvement Notices declined in correlation with the increase in the use of the FFI facility. Initially, the only route of appeal was to the enforcement agency itself, although following a massive groundswell of opinion against this flawed approach and mass resignations by the more honourable inspectors, an independent board of arbitrators was established.

So, it is not just Spain that produces such irony but my own homeland and probably many other countries. However, I will stick with Spain just a little longer, as during my five/six years of residence, I saw the other side of health and safety. As a serving magistrate though, I was fully aware that there are always three sides, at least, to every story. The version put forward by the prosecution, the version put forward by the defence and the version decided upon as fact by the magistrates. This was to support my approach to following a pragmatic view of health and safety to this day.

Sometimes, the facts put forward by the prosecution are overwhelming and incapable of reasonable rebuttal. At other times, the prosecution cannot even get past the 'halfway' stage following a successful submission of 'no case to answer' by the defence.

On one occasion, however, shortly after moving to Spain, while in temporary accommodation, I was fortunate, or unfortunate, to witness what I believed to be overwhelming evidence of total incompetence. A spare plot of land adjacent to where I was living was purchased for what I was to find out was the development of a block of flats. In the UK, all the neighbours would have been informed of the proposed project. In Spain, it just happened. Anyway, under normal circumstances this would have been a straightforward construction project but for the fact that it was immediately adjacent to the road that ran along the shoreline.

I watched the demolition of the pre-existing derelict bungalow with horror, as it just looked like a bunch of mad men (no women) being let loose with sledgehammers and a wrecking ball to bring it down. Having said that though, there were, miraculously, no injuries and it was all done and dusted in a day. That is not strictly true, as the dust lingered in the air for days.

The next day the diggers turned up and again I watched with horror as none of the expected safety features were implemented, not even a barrier to protect the public. They were effectively walking through the site, oblivious to the risks. No protective safety equipment was being worn by any of the construction workers other than the customary impact-resistant sun hats and steel toe-capped flip flops. It got to the stage where I dreaded looking out of the window but for the sheer intrigue of what they could possibly get up to next.

I was not to be disappointed. The first concrete pad was being formed, approximately 6m below ground level. I had already started to wonder if any shoring up of the sides was to take place, though it should have been their very first task. You see, unless you are already a step ahead of me, this development was being constructed immediately opposite the shoreline to the Mediterranean Sea, below the water table.

I could contain myself no longer. I attempted to speak in Spanglish (a weird concoction of what I knew at the time of Spanish with English language infilling) with the person who looked as if he was in charge. As you can imagine, this very strained dialogue did not get very far, with much shrugging of the shoulders. I could not leave it there so arranged for my Spanish-German co-director, who obviously knew the lingo and the system, to take the matter further. Who knows, we might even have obtained some consultancy work.

After about an hour trying to explain the problem with constructing anything below the water table, he reluctantly agreed to take it further. I am still not sure that he fully understood the problem. Anyway, eventually he came back to me with the suggestion that we mind our own business as the project did not need any advice and all was in order, as all the necessary paperwork (toilet paper) was in place. Coincidentally, the site had already started to flood that morning.

I then sat back and watched the arrival of the pumps and shoring equipment before the actual work recommenced about a month later. I just wish smartphones had been around back then as the rest of the construction work presented a masterclass in how to do it wrong. Even if it were to have been fully choreographed, I would not have dared produce such dangerous practices.

I will leave you to use your own imagination when I mention the exterior painting, using upturned beer crates to get to the underside of the balconies above. There was no edge protection at all and no fall restraint or arrest, until a cradle was rigged to paint the side walls. Yes, the painters in the cradle were wearing harnesses but again that tuning fork started vibrating. Something was not quite right. I was transfixed but all became clear when they lowered themselves to my eyeline. Yet again, I could not believe what I saw. Where would you think they were attached? Well, they should have been attached to the anchor point in the base of the cradle. No. They might have been attached to an anchor point on the building, even though that would have been a dangerous practice. Again, no. Have you guessed yet? Try each other. One out, all out.

This debacle was, effectively, to have been one of my last actions in the wonderful land called Spain and I sincerely mean that as a place to live and relax.

I had no choice but to stop the money being drained out of my UK company. I terminated all contracts and handed back the keys to the nicely refurbished medical centre, offices and safety training facility in the heart of Málaga. I was not to know until I returned to England that the International Guarantee had also been called in and my UK company was required to repay, on demand, the full value. All in all, the venture cost me both of my companies, about £0.25m in cash, and eventually my marriage.

However, although the lessons learned were costly, in so many senses they were worthwhile and gave me the insight to apply hard-earned wisdom. Sometimes, we get through life on just pure luck but generally we are blessed with the ability for self-preservation. Even stupid people possess this and whether by a miracle, a quirk of fate or something else inside us all, humanity has survived through thick and thin. Yes, I rebuilt my company and yes, I put into practice the acquired knowledge and the belief that we really do not need to be told, as adults, how to protect ourselves. The laws that we all now must abide by have only evolved to protect us from the idiots, the uncaring and the unscrupulous.

CHAPTER 3

SELF-PROTECTION

Just stop and use your senses

Even the grandfathers I mentioned in the first chapter and those Spanish workers in the last one had the ability to protect themselves instinctively. It was probably why most of them escaped with their life and all their bits intact at the end of their chosen careers. From the cradle of humanity, we have all been able to do this and some are obviously better at it than others, or just extremely lucky.

Let me try and explain. Once you have read what I am going to ask you to do next, you might want to close your eyes and think of yourself in the moment. You will need to imagine that you are one of your ancient ancestors. It is thousands of years ago and you are sitting by the fire in the mouth of your cave. You are waiting for the sun to rise so that you can go out hunting or gathering. Akin to going out to work today, just a bit more primitive. You utter some sort of words to your chosen partner and, if you have them yet, your offspring. You are just about to leave, and wow...

You get a funny feeling inside. It is a bit like a tuning fork but very strong. You stop dead in your tracks. You know something is there but you have no idea what that something could be. It might be a sabre-tooth tiger or a bear or anything really. You just have a feeling, the tuning fork inside you is ringing like crazy. Like an alarm gong now.

I will come back to this tuning fork many times throughout the book. However, at this point please now close your eyes and humour me with that very short exercise.

In our developed world, you will not have to worry too often about being mauled to death by that tiger or bear. That does not mean those same ancient neural mechanisms are not very much around today. You must have experienced the feeling of your own tuning fork and I just want you to remember what it was like. Cast your mind back. You walked into a room or a space and thought, wow, I know something is here, or somewhere close by, but just have no clue what it might be.

Let me explain to you what nearly all animals with a backbone have possessed, to protect themselves, for many thousands of years. It is a wonder hormone called adrenaline. It is a powerful little molecule that can change your personality from a logical human one second to a true survivor the next. It can make you more anxious or ready to jump out of a plane, and when you feel threatened it is your one last safety net.

Without it, none of us would be here today.

Freeze mode (take 2)

Hopefully, you will have remembered the sensation of your tuning fork, probably more than once or twice. If you prefer, you can call it your alarm bell, but for consistency I will refer to it as a tuning fork, as sometimes it can go unheard, whereas an alarm bell cannot normally be ignored.

At the point where our tuning fork activates, most of us would stop what we were doing and try to work out what it was. This is the very first survival instinct. When you feel threatened, you will normally freeze, if only for a split second and your adrenal gland will start pumping adrenaline. It is how your body gets ready to react. In that split second, your heart rate will increase, your blood pressure will rise and you will start to sweat. These are all signs of what is classically known as the fight or flight response. I will come to this soon enough but it is worth staying with the subject of freeze mode just a little longer.

Unfortunately, as I mentioned just before, we have grown accustomed to not having to worry about being mauled to death by that tiger or bear. In fact, I would go as far as to say that we have generally become quite complacent about the hazards (anything that has the potential to cause harm) and risks in the home, our workplace and our outside activities and pursuits. Remember those grandfather rights, where shortcuts are taught without the underpinning knowledge of the reason for doing things the right way. Also remember those customs and practices that give you a false sense of security that your boss is responsible for looking after you in the workplace.

So, let us consider what could happen when your tuning fork activates but you think to yourself you have always done it that way, or your workmate has your back, or anything other than listening to your alarm bell. It is not quite loud enough to make you stop. Well, go on, guess what could happen next.

The next thing you know is, if you are one of those lucky ones and after you are brought out of your induced coma, thinking, why did I not listen to that damned tuning fork after all? Why did I walk headlong into that danger that I would have seen had I just looked? Why did I not react to the noise of that whining motor? Why did I not just turn around and walk out when I smelled that funny smell? Why did I not wear that PPE that I had been told to so many times? Well, it could be all those things, as the tuning fork kicks in to override your normal behaviour for a reason. You failed to react to it at all. That is what happened next.

That is the reason why I am sticking with the freeze mode for a little longer.

So, instead of saying "if only" after the event, the next time you feel that vibration, ask yourself "what if" my inbuilt warning system is telling me that there is something around that is going to cause me harm.

To do this you will need to take a step back, either literally or figuratively, to give yourself enough time to understand what your senses are telling you about the danger before deciding on whether fight or flight is the best option. If you do not override your tuning fork this does not have to take very long and in fact, when the danger is imminent, this can occur in a split second.

In workplace settings, however, you would do well to 'take 2', which is a simple system that I have recommended to all my clients. It represents just two minutes of your time before starting a new task and simply involves weighing up the situation rather than just bowling in feet first. As an *aide-memoire*, you might like to consider producing a small pocketbook of potential hazards. I do realise that your ancestors from chapter 1 did not have such a luxury but we have moved on and although it remains the 'common sense' thing to do, as I am sure you will understand by now, 'common sense' is not that common.

Let us 'take 2' and return to chapter 1 quickly. Hopefully, you will recall the criteria for 'common sense' to be a valid tool, as laid down by Sir Francis Galton. However, in this instance you will not have a mechanism to collate everyone's independent ideas so you will have no choice but to rely on your senses alone. Instant decision making is not uncommon though and is based on many things, including gut instinct, but I am going to suggest something that indirectly came out of the findings of a fatal accident investigation.

I was a part of an investigation team put together in July 1991 following the deaths of two firefighters in London. This investigation was the precursor to the issuing by the Health and Safety Executive (HSE) of two Improvement Notices. One of these concerned the operational training standards and from this a process of dynamic risk assessment (DRA) was developed for the UK Fire and Rescue Service, which has since been adopted worldwide.

Prior to formal DRA training, officers would always weigh up the situation before committing crews. The process started at the fire station when the call came in and basic information was provided as to type of incident, location and resources ordered. This information would be supplemented on route to the incident, from radio traffic, other members of the crew or visual information prior to arrival. The most important intake of information was imparted to the officer in charge of the first fire engine to arrive from what she/he saw playing out in front of the crew.

It is difficult to put into words the sheer quantity and diversity of information that had to be assimilated in a very short time by that first officer and the need for early and correct deployment of resources. This sort of DRA process will, of course, be understood by all blue light and military personnel but is not a daily occurrence in commerce and industry. As I suggested previously, this was always undertaken instinctively but in 1991 the process was put under the microscope.

As you will imagine surely, much more than a simple 'take 2' procedure was developed but I am going to stick with that for the purposes of encapsulating the freeze mode. DRA is simply the practice of mentally observing, assessing and analysing a working environment, to identify and either remove or mitigate risk. It allows individuals to identify hazards on the spot and make quick decisions about their own safety. You will do these thousands, if not millions, of times each and every day of your life without a problem, but in the workplace, employers have a 'duty of care' (see chapter 5) and need to be able to prove, after the event, that an operation had been assessed prior to starting.

That is difficult to do when instant decisions are made on the hoof. There needs to be a simple process that facilitates this evidence gathering and that does not cause too much of a delay. Hence, just a two-minute pause should be sufficient. You will understand, I am sure, that even this delay would be unacceptable on the fireground or a battlefield but not so in the factory or a warehouse. Maintenance engineers use DRA routinely but few of them realise it at the time. To 'take 2' minutes before climbing a ladder or stopping a production line to release a blockage is all you need to prevent falling from height or crushing your fingers.

In non-emergency situations, that *aide-memoire* I mentioned could be used to good effect by recording some basic information regarding the operation, the equipment and the environment etc. This would both focus the mind on the hazards and provide that vital evidence should it ever be needed.

So, the next time you feel your tuning fork vibrate, just 'take 2' and understand why your senses have triggered your adrenal gland. Understand also that the adrenaline being produced is flowing to your brain and muscles in readiness for you to either take on that sabre-tooth tiger or run away from the bear.

Fight mode (normal procedures)

As hunters and gatherers, we have learned many different skills, from ancient times and in modern society, that we put to good use every day. Men and women everywhere are going to work, effectively taking on that sabre-tooth tiger, gathering food and/or caring for others. Many of these activities have been refined into raising livestock and farming, to working in a factory, construction work, driving vehicles and flying aircraft, even space exploration, but essentially they are all based on standard working procedures for a particular work activity. Yes, of course these are developing all the time but essentially the fight mode is now enshrined in a set of routines that we hope are safe. None of us want to go and fight the sabre-tooth tiger and be eaten by it, which is why our ancestors became so good at tracking, camouflaging, trapping, killing and skinning these animals for either food or clothing.

These became routine tasks, where fathers taught sons and mothers taught daughters the way to do things properly, without getting hurt. Although gender is no longer a prerequisite of the ability to go to work, in most modern countries routine and regular tasks still present risks, a subject about which I will discuss in greater depth in chapter 7. For the purposes of this chapter, however, all you need to remember is that we humans generally prefer the comfort of knowing that what we are doing is either inherently safe, or if not, we are in control of the dangers. For me, I loved being in control of the danger, whether it be firefighting, jumping out of perfectly serviceable aircraft or riding a fast motorbike.

When you are operating outside of your comfort zone, you will experience fear and without fear you would be a danger to yourself and others. Fear is a good thing, as it will guide you through your fight or flight responses and help you to keep safe and alive. The negative side of fear is when it holds you back from doing something positive. Either way, you will want to overcome your fear(s) and the only way to do that is by taking the appropriate action(s). This is all part of your adventure called life and is so vital when it comes to your survival. Fear will keep you alert by heightening your senses and awareness. As we have already discussed, it triggers the adrenaline flow so that your brain can assimilate what it is being told by your senses. That adrenaline also primes your muscles to react appropriately, in accordance with your skills, knowledge and experience.

In a modern workplace, no matter what the activity involves, a standard operating procedure (SOP) is required to ensure uniformity of working methods. These SOPs will be based on all the hazards involved and those that can be foreseen. Their development will entail the identification of all the things that can go wrong with both the regular work activity as well as any infrequent or unusual tasks. They are step-by-step instructions compiled by an organisation to help its workers carry out routine operations. They should always aim to achieve the highest levels of efficiency, effectiveness and safety.

For some workers though, routine tasks are not routine, such as the military and blue light services and these organisations might use the term standing orders instead. These are not standard, as although fireground or battlefield operations are inherently the same, each incident or battle is unique. In these circumstances, the operational tactics might differ but strategically they all require the same quality of output (ie to put the fire out or to win the battle) and general uniformity of performance, while reducing the potential for failure when it comes to communication and leadership. Regulations, statutory guidance and accepted good practice will also usually feature.

So, let us return to our cave dwellers for a moment. What sort of SOPs do you think they would have developed?

Well, albeit much less sophisticated, in taking on a wild animal for instance, they would have needed to plan their method of attack. The preference would have been to do it as a family unit. All members of the family would need to be fit enough to undertake the task. Those with the most experience would have trained the others. The method used would have been efficient, effective and, although dangerous, safe. Sometimes, a family member would be killed or maimed, but had the method not been safe overall, once again we would not be here today as a species.

I will leave you to fill in the gaps, but needless to say, their SOPs worked and we will see just how, fundamentally, today.

Flight mode (when things go wrong)

Oh no. That sabre-tooth tiger you snared last night has heard you coming. It is angry and very hungry. It just wants to kill and eat you. As you approach, it breaks loose and in that split second you become the hunted. The adrenaline is really flowing now. You had no time to think, you just turned and ran. Your leg muscles are fully pumped up and taking you somewhere just as fast as they will carry you. The trouble is, you just do not know where right now. You must make some instant, life-saving decisions on the run, literally. Hold on, you can see a tree and even though you are more than capable of climbing it, you know that the wild animal chasing you is a good tree-climber too. Your brain is in overdrive now. You know the sabre-tooth is a much faster runner than you but what else can you do?

In an instant, your brain tells you that at least you might have a chance to fight it off from above. You have no choice. This is life or death. You have given yourself a two-second advantage. Is it enough?

Hold on. You forgot you were part of a team. This had been planned for and two seconds might just be enough. Just behind the fast-approaching tiger is the rest of your family, with spears and clubs. You only need to hang on for a few more seconds. A lifetime. You get into the tree but you just do not have enough time to get any higher. You have one last chance and, luckily for you, you are still carrying your club. Your muscles contracted so tightly you just kept hold of it and now you must use all your skill to repel the beast. One blow on the snout was enough to get it to falter and that was just enough time for the rest of the family, collectively, to subdue your attacker.

You can now breathe a sigh of relief and climb down. Wow, that was close. You are exhausted and that adrenaline is still flowing. You can still feel your heart pumping so what did it feel like to be the prey? What did you learn from your experience? Probably, the next time you build a snare you will do a better job.

Strangely enough, the analogy of being chased by a beast reminds me of the experience of fighting a large fire. I was always taught that fire acts like a beast being in a rage and totally unpredictable. If you can remember back to your science lessons and the triangle of fire, I am sure you will recall the three ways of extinguishing a fire. One way is to take the heat out of it, which is why all fire engines carry large quantities of water. However, a bit difficult to either drown a raging sabre-tooth or throw it in the Arctic Ocean so it dies of hyperthermia. Another way is to starve it of oxygen, but again, you would need a very large pillow to suffocate this beast and finally starve it of its fuel, or food in the case of the tiger. Again, you did not have that long. Much better to hit it with your club and spear it to death, but either way, you managed to overcome your fear to fight another day.

In the heat of a dangerous or catastrophic event, especially in the very first few seconds and minutes, you will be in pure survival mode. However, we have all been gifted with a survival instinct and mechanism to respond and save ourselves from disaster. Sometimes, it plays out in our favour and sometimes it fails us, but survive or fail as individuals, we do continue to survive as a species.

Over time, we learn from these events and develop an emergency response procedure (ERP) for each eventuality. Even so-called 'acts of God', like floods, volcanic eruptions, earthquakes and tsunamis, are known to us as humans and we have established and developed mechanisms for dealing with them. There are also more localised catastrophic events, usually occurring through some form of human error, like fire, collapse of a structure and even intentional destruction from such things as riots and terrorist activity.

These known incident types are generally handled in similar ways around the globe but with local variations. You will be fully aware, I am sure, of what to do if you discover a fire in the workplace, how to raise the alarm for others, and if you have been trained and feel confident, how to extinguish a small fire before getting out of the building and leaving it to professional firefighters. It is also important to practise these ERPs in order that everyone knows the 'drill' and can respond correctly should the real thing happen. Unfortunately, though, if these drills happen too often, you and your work colleagues are likely to become complacent, and if they are then not undertaken correctly, these actions will not help you when needed and you will be left to rely on those inherent survival instincts.

For instance, how many times has the fire warning signal activated for you to simply leave the building, answer your name in a roll call and walk back in and carry on working? Ask yourself what good did that do and, indeed, did it even turn you off to the whole idea of why it was needed in the first place? Would it not have been more useful had there been some realism?

In future, if it is up to you to plan such a drill, you might like to consider planning for it to go wrong. Perhaps ask one of the fire marshals, or whatever you might call them, even someone who naturally takes a leadership role, to pretend to fall over on the staircase. This should then force other members of the workforce to think for themselves. It will make the whole exercise more memorable and facilitate a more worthwhile debrief as, surely, that is the whole point – to learn from our mistakes – isn't it?

You have a choice

Sadly, there are those among us who make the wrong choices. They might be selfish, lacking consideration for their fellow worker or anyone other than themselves. They might be too busy to look after someone else, being wrapped up in their own world. These and people like them tend to just 'walk on by'. Even worse are those who intentionally get in the way and/or actively disrupt things, whether that be the SOP or ERP.

I call these people 'terrorists' in the figurative sense but hopefully these people are in the minority. You should always seek these people out, especially during training and practice sessions. They might just fail to respond. They might just moan a lot. They could actively disrupt the proceedings. Either way, they need to be told the error of their ways, brought into line or, as a last resort, dismissed.

However, these 'terrorists' might have seen the futility of the process but just can't or won't pass on their observations. They could provide you with valuable first-hand experience of a procedural error or unnecessary element but no one seems to care. They might have a pearl of wisdom to share with you that will make the whole process more efficient, more effective and/or safer. If only you would ask and then listen.

Sometimes, people get so wrapped up in what they are doing that they become oblivious to what is going on around them. I was initially shocked to learn that some of the most prolific victims of falls from height were electricians. This felt counterintuitive to me. When you consider their primary function, I am not sure if you will agree with me in believing that electrocution would have been the predominant type of accident.

However, think again about where they might be required to perform their work. Think also about the intricacy of that work. Consider what might be occurring in the electrician's mind while wiring a distribution board at height. Until relatively recently, she/he would probably have been using a ladder, although in the UK ladders should now be reserved for access only, or short duration (up to five minutes) and low-risk activities. From an analysis of incident reporting though, relating to falls from height, many of the victims have disclosed that they simply forgot they were on a ladder or working at height.

I attended one such incident, in the City of London, where an electrician was performing such a task and simply forgot he was on a stepladder at all. He simply turned to walk away and fell. He was not even that high, probably about three steps up, so about 1m. What he failed to recall though was that he was working on a construction site. Normally, he might have just twisted his ankle or stumbled and regained his footing. On this occasion, he was not to be that lucky. Sticking out of the wall was some unprotected reinforcing bar. He fell straight on to it and was impaled. My team and I managed to cut him away. He survived but this was a life-threatening incident. Had he chosen to work from inside a podium, it would not have happened in the first place.

Although both the electrician and the 'terrorist', figuratively speaking, were acting subconsciously, there are also those who make conscious decisions either not to help others or even to intentionally harm them. I am not talking here about the individual that protects himself or herself at the expense of someone else but the prankster. These people need to be identified early and removed, as they are positively dangerous.

My time in the Fire and Rescue Service saw many practical jokes and many were ingenious and a great deal of fun. It was a part of the culture and I could even see its purpose initially in developing a team spirit and camaraderie but there was a flip side. Firstly, it was a form of bullying and for that reason alone is not welcome in the workplace. However, it was also potentially very dangerous. I recall one very sad incident that changed my mind completely.

First, I need to describe how a piece of furniture was built, as that was the essence of the prank. The cushion of a mess room chair was supported by removable webbing straps. It was custom and practice at one of the fire stations in central London, where I was the officer in charge, to play a prank by removing these straps, so that the unwitting member of the team would fall through when she/he sat in the chair. This would cause immense hilarity, at their expense. It was all taken in good part and no one was ever hurt except for their pride, but this was all about to go horribly wrong.

On this fateful day, the prank was played out on a new member of the team (called a watch). He had just been posted to his first station, to my watch, after 13 weeks at the firefighter training centre. He would probably never have been any fitter. He had always wanted to be a firefighter and was so keen. Like all those who had gone before him, he duly sat in the seat that was earmarked for him and fell through, but this time it did not happen as planned.

The cushion did not fully come away to protect him, as it always had done. The expectation was that the cushion would hit the floor first but his full weight landed on the solid floor. It was not even carpeted. I was not in the mess room at the time but I could hear his pained cry from my office. He had broken his coccyx, not that we knew it at the time. Everyone rallied round but there was little to be done other than getting him out of the chair and making him comfortable. We awaited the arrival of the ambulance. Suddenly, it was no longer funny.

That was not the end of it though, as the joke was also about to backfire further. The injury turned out to be life-changing for the poor young man. He was placed on sick leave and after many weeks it transpired that he could not return to operational duties. At that time, it was not possible for active firefighters to be transferred to non-operational roles so he was medically discharged. He reapplied for a non-operational position and was successful but it was only ever a desk job. Effectively, he had not even started his chosen career.

The team were devastated. The young man lost his job. Was it worth it?

CHAPTER 4

MAKING MISTAKES IS NORMAL

We all make mistakes

Cast your mind back to the 'flight mode' scenario, where I suggested that you plan for things to go wrong in a fire drill. I am going to use that analogy to support my point about needing to make mistakes, hopefully in a safe environment, if we are to learn from them. In planning for things to go wrong safely, you will also notice, after the event, that the participants will remember both the scenario and the outcomes, much more than if they just go through the motions.

This is particularly so when you are the person confronted with the task of instant decision making and even more so when it is outside of your comfort zone, say a real fire event. If you keep doing the same thing and never get challenged, apart from it being boring you will never learn from the little mistakes. These little mistakes can also act as a sort of warning signal, which is not to say that they replace the tuning fork, just save you having to rely on it for when the real incident occurs.

In fact, back in 1931, Herbert William Heinrich developed a theory of accident prevention, although it has been further developed since. He was an assistant superintendent in the engineering and inspection division of an American insurance company. After reviewing 75,000 injury and illness cases, he showed a relationship between serious accidents, minor accidents and the underlying unsafe acts and/or conditions (known as a near miss, or more accurately a near hit). From 12,000 insurance records and actuarial reports and 63,000 reports from industrial engineers and plant managers, he found that for every accident that causes a major injury there were 29 accidents that caused minor injuries and 300 accidents that caused no injuries (the near miss). This was probably the foundation of behavioural safety and is known as Heinrich's Law.

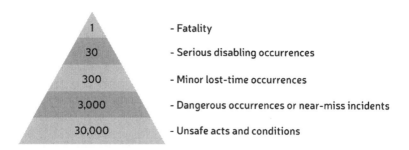

The Heinrich Pyramid was updated, notably by Frank Bird in 1966, based on an analysis of nearly two (1.7) million accident reports and others, by extrapolating and refining the original pyramid, so that for every single fatality, there would be 30 serious disabling injuries, 300 minor lost time injury incidents, 3,000 non-injury or dangerous occurrence (near miss) incidents and in excess of 30,000 unsafe acts and conditions (ie hazards, which if you recall are anything that can cause harm).

Even though both pyramid portrayals have been criticised for being relevant only to those industries with large workforces and the ratios changed between different types of activity and industry sector, I believe you will concur that they still paint an overall picture that accidents are both predictable and preventable. Criticism has also been levelled on the focus on the reduction of minor injuries but I do not think that is either true or a fair assessment either. The purpose of the general philosophy is simply to take appropriate intervention early on and this can positively affect everything that comes after. In other words, address the underlying hazards and the risk(s) will not materialise.

For some reason, several academics have tried to disprove this basic philosophy by suggesting that the 'root' cause of all accidents is not related to human action(s) or inaction(s) but rather poor management systems. Once again, I believe that if we seek the 'common sense' of the wider population this contention will also fail, as behind all management systems are humans. It could be poor system design but that is human error. It could be poor system operation and that too is human error. It could be poor data input or interpretation and that is also human error.

The bottom line here, in my opinion, as a qualified and experienced incident investigator, is that human error is an element of all accidents. It is correct though that a thorough investigation should never stop at that point, and if you are ever called upon to carry out an incident investigation, you should always keep asking one critical question. Why?

We are human and we make mistakes. If incident investigators, qualified or otherwise, stop at the level of the individual(s) who made the mistake(s), that would be simply laying blame and that will not prevent someone else making the same mistake. By asking a series of 'why' questions, you will identify the correctable system failures, which are most often management failures. For instance, all system failures will come back to things like poor leadership or decision making, unclear responsibilities or poor communications, lack of proper information or training, lack of proper planning, poor system or equipment design, unsafe customs and practices, poorly maintained or unsuitable equipment, poor supervision, or sometimes, unfortunately, things like horseplay or bullying. All these root causes are the symptoms of failure of management control.

Learning from mistakes

So, it would be beneficial to all of us, at home, at play and at work if we could stop the serious consequences of our human failings by analysing the events and conditions lower down the safety pyramid, learning from those mistakes and putting them right. You will not need to consider undertaking a formal investigation or analysis at home or at play but the opportunity to learn and act is still there. The 'common sense' tends to arise from the 'common mistakes' and invariably, the root causes will repeat themselves throughout your life. Most are totally foreseeable and avoidable so why do we not foresee or avoid them?

However, sometimes catastrophes are unforeseen, such as with the disastrous fire at King's Cross railway station in Euston, London on 18 November 1987, where 31 people died, including one of my colleagues, station officer Colin Townsley. Sixty more people were injured.

The now understood 'trench' effect that caught out Colin and the others who died with him was a totally unknown phenomenon at the time and lessons had to be learned to ensure that this never happened again. As the station commander at Euston fire station, I was also called upon to attend the public inquiry and listen to the evidence, which included the rebuilding of a scaled version of the scene of the incident and a full video capture of this then new effect. It was obvious, in hindsight, that this was primarily caused by a small fire underneath the wooden escalators that developed unseen or undetected until it broke out, and was caught in an air current that was channelled upwards by the side panels of the escalator, like a blowtorch.

As with most major catastrophes, the background and history leading up to the event was also instrumental, and here the unsafe acts and conditions could and should have been foreseen, had anyone looked. The accumulation of rubbish under the wooden escalators provided the fuel. That was made worse by the grease from the running gear, due to poor maintenance standards. At that time, smoking was allowed on the underground, which it turned out provided the ignition source, and there was a plentiful supply of air (oxygen), especially when the trains were entering the station.

We have, therefore, a triangle this time, the Fire Triangle, which as you probably know, contains the constituent elements needed for fire initiation and growth, being the fuel, the ignition source and the oxygen supply. This part of the story was completely foreseeable. The common sense would have been to remove both the fuel and ignition source well before the event happened. Sadly, there had been several 'near miss' events previously but few people took any notice. Although very difficult, virtually impossible, to do anything about the third part of the triangle, the oxygen supply, removal of the first two pre-incident unsafe acts and conditions would have prevented the inevitable disaster.

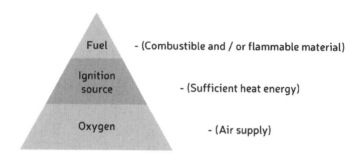

Fuel - (Combustible and / or flammable material)

Ignition source - (Sufficient heat energy)

Oxygen - (Air supply)

In fact, the chairman of the public inquiry, Desmond Fennell OBE QC, who was assisted by a panel of four expert advisers, criticised London Underground severely for its poor attitude towards fire safety. It found that members of staff, at all levels, were complacent about the whole issue, as there had never been a fatal fire on the underground. They had been given little or no training to deal with fires or the mass evacuation of members of the public. The inquiry team recommended and London Underground implemented a wide range of actions which were considered necessary to prevent a similar disaster.

There were 157 prioritised recommendations in the report, including the fitting of closed-circuit television (CCTV) equipment, public address systems and automatic fire detection throughout and fire suppression systems in the machine rooms under the escalators, alongside a reinforcement of a smoking ban that had come into force about five years earlier but was being widely ignored, with a total ban on the sale of smokers' materials, both below and above ground at stations. All wooden parts of the escalators had to be replaced with metal and the station supervisors had to personally inspect the escalators every two hours until all wooden parts had been removed. Cleaning had to be stepped up, along with maintenance.

Management issues featured highly, as you would probably expect, and focused primarily on senior leadership, or lack of it, especially at director level. Responsibility for ensuring that formal and strategic safety monitoring occurred was allocated to a non-executive director, reporting directly to the board, as well as worker (Trades Union) participation at safety committees. All managers were required to undertake refresher training every two years and other members of staff every six months.

The emergency services had to be provided with up-to-date station plans, the Fire and Rescue Service had to be consulted when station alterations were planned, and legislation was enacted by the government in relation to all sub-surface railways, which included enhanced responsibilities for the Railways Inspectorate.

Many other recommendations were made, including those to relieve overcrowding and some structural alterations. It took many years to implement the recommendations but nothing like this has been seen, in the UK at least, since and the terrible lesson was, indeed, learned but, wow, at an enormous cost. Had those little things, which were, remember, foreseeable, been addressed early, the disaster and unnecessary loss of life would have been averted.

You must be wondering just how many other foreseeable disasters are waiting for us around the corner?

Instead of saying "if only" ask yourself "what if".

The playing cards are being turned over all the time and the joker is still in the pack.

Do not wait for the joker

The only way to stack the odds of the joker being turned over is to increase the number of cards in the pack. The equivalent in terms of accident prevention is to look at all those unsafe acts and unsafe conditions. You need to start looking or intensify your search. Make sure your team members open their eyes to spot the clues too. If you look, you will find that they are all hidden in plain sight. In the build-up to every fatal accident there are, statistically, 30,000 clues at least. There are 10 clues at least in the build-up to a near miss, so if you look, you will find them.

Even when it comes to 'natural' disasters, the human species has developed plenty of warning devices. Even so, they continue to cause hundreds and sometimes thousands of deaths. Take the Covid-19 pandemic as one example, that caused millions of deaths and economic destruction throughout 2020 and 2021.

Without getting involved in the politics, only because it would take us around in circles, scientists knew about the coronavirus and its effects. Somehow, it escaped its controlled environment and spread so fast that it took the world by surprise. The rest is now history but many political, medical, economic and scientific questions still need to be answered.

- You must be wondering, again, was this yet another foreseeable disaster?

- Was it preventable or could it have been stopped in its tracks, as some countries managed to do?

- Were we all let down by our leaders?

- Will the world be better prepared for the next pandemic which is, surely, just a matter of time?

- How will the diversion of healthcare resources to manage the current and future pandemics affect the ability of healthcare systems around the world to maintain standards of care for those patients with other life-threatening medical conditions?

- Has the arrival of ever more virulent and vaccine-resistant strains of infection and disease undermined our ability to control new outbreaks?

- Will the global economy have the resilience to concurrently develop worldwide infrastructure and support countries recovering from the depths of a pandemic crisis, while continuing to provide debt relief for the poorest?

- In a world where humans are increasingly connected and moving around the planet faster than at any other time in history, should we expect to see increasingly more robust cross-border restrictions?

- What effect will climate change have, moving forward, on fuelling the spread of pandemic disease?

What else is lurking in the wings though that, if those responsible were to look, has the potential to cause harm? This is the classic definition of a hazard, and hazard identification is the first step in determining the existence, nature and severity of any risk, let alone assessing how it can be controlled or mitigated. The only tool necessary in undertaking this most basic of human activities is to first look for it, although you should be using all your senses, including intuition. As the common phrase goes, "Stop walking around with your eyes closed." This is still only half the story though, as once you have found that something can harm you or someone else, would you just walk on by?

I recall a story about Ken Woodward, a factory worker at a Coca-Cola plant, who was blinded in a chemical incident. He went on to tell his story very publicly and it is well worth checking out in terms of behavioural safety and the pressure from the workforce to improve the safety culture of the organisation. There was one part of the story that really captured my attention, which concerned one of the maintenance staff who went above and beyond.

I seem to recall that the incident happened over the weekend and on the Friday before, one of the maintenance operatives saw, as he was leaving work for the day, that the safety shower was faulty. It troubled him to the point that he went back to work first thing the next day, in his own time, to make sure it was working properly. It did not stop the incident occurring and did not stop Ken being blinded but it did save his life.

He looked, took notice and acted. That is all it takes.

The tell-tale signs

So, hopefully, you are starting to see a picture emerging that there is a 'domino' effect occurring in all accident sequences, with a myriad of interconnecting small signs. Some of the signs are big, or at least in plain view, for all to see and I wonder sometimes just how they can be missed. It begs the question, who is ultimately responsible for these 'failures' and surely it has to be the person at the top, doesn't it?

It might well be the leader turning that blind eye again but every single member of staff can do the right thing. Acceptance of the culture of turning a blind eye is a corporate leadership issue but every single one of us, you included, has a moral obligation to look after ourselves, our nearest and dearest, our friends and our work colleagues. Yes, it is good to make the little mistakes but it is not good to ignore them, as they grow into bigger ones and contribute to the ultimate catastrophic event.

This is so evident after the effect, when major disasters are analysed under the microscope, from King's Cross as I have already discussed, to the Bhopal (India) gas leak in 1984, which killed between 3,787 (government figures) and 8,000 (unofficial records) predominantly innocent people, to the Deepwater Horizon (Gulf of Mexico) oil rig fire in 2010. In the case of Deepwater Horizon, which repeats so many others, the inquiry found a whole catalogue of failure events, where each one could have been addressed positively as an individual issue to prevent the next, and ultimately the final, devastating explosion, which miraculously only killed 11 people and injured 17 others. However, the environmental damage was enormous, killing thousands of marine mammals and sea turtles, as well contaminating their habitats.

I have also analysed a similar oil rig fire as part of my specialist fire protection training, being the Piper Alpha (North Sea) oil and gas rig explosion and fire in 1988 which killed 167 workers; 61 workers escaped and survived but 30 bodies were never recovered. The total insured loss was about £1.7 billion ($2.3 billion), making it one of the costliest man-made catastrophes ever at that time. This was to be totally eclipsed by the Deepwater Horizon costs which reached $65 billion.

In short, the Piper Alpha disaster demonstrated the 'domino' theory perfectly, as the original explosion was not the one that did the ultimate damage. A system of permitting was being used, which was not being coordinated. One permit had been issued to work on a gas pipeline, with a fixed finish date and time, to ensure that the system remained isolated while the work continued. During the period of the first permit, another permit was issued to undertake some hot cutting in the same area. I am sure you have already guessed that the two things together are just asking for trouble.

Yes, you are correct, the inevitable explosion occurred but that was not the one that led to the final death toll. In the scheme of what was to come, it was a relatively minor explosion that could have and should have been contained. It was just the start. A whole string of other unsafe acts and unsafe conditions just compounded the problem. A failure of the fire suppression system did not help. This was a historical problem that had been identified many times. Are you starting to see the common trend?

I am not going to go into the detail of the inquiry other than to say that history repeats itself, over and over and over again. All those little things that should have been addressed by everyone involved were not, and eventually the fire ruptured the main oil pipeline for the rig and the rest, yet again, as they say, is history.

Leadership failure

In every single public inquiry following a disaster, whether natural or man-made, the same basic failures are 'discovered'. A failure of leadership is invariably in the mix, together with poor communication and system failures, which are management failures in any event. The final reports all mention the inability to learn from previous mistakes.

Perhaps you saw the film about the Chernobyl (Ukraine) nuclear reactor accident in 1986. The cause was put down to design flaws in the reactor and serious breaches of protocol during, ironically, a simulated power outage safety test. The manager of the plant would, quite simply, not believe the information that he was being given. Not only was this information coming from the main control panels but it was also being corroborated by first-hand accounts of what was happening inside the reactor chamber itself.

The system of state control that was being imposed on the manager by his political masters was so great that he feared any form of failure and simply explained to them that everything was good. It definitely was not good. As the incident escalated, he refused to do the things that he should have done, including shutting down the reactor, for fear of recrimination. All opportunities to prevent the ultimate disaster were missed.

As official figures are contested, the consensus is that a total of approximately 30 men died from immediate blast trauma and acute radiation sickness (syndrome) in the seconds to months after the disaster, respectively. Another 60 are believed to have died in the decades since, including those with radiation-induced cancer. However, there is still disagreement concerning the accurate number of projected deaths due to the disaster's long-term health effects. These range from up to 4,000 for those most exposed – the people of Ukraine, Belarus and Russia – to 16,000 in total for all those exposed on the entire continent of Europe, with figures as high as 60,000 when including the relatively minor effects around the globe. The environmental impact was devastating and exists, to this day, with ongoing and profound psychological impact, including high levels of alcohol and drug abuse and suicides, with the economic damage caused by the disaster estimated to be in the region of $235 billion.

So, if leadership, or lack of it, is always in the mix, maybe you should ask yourself a few questions at this stage (more about leadership later).

- Do you have vision? Good leaders will want to inspire their team(s) to go on to bigger and better things. They will want to see the team(s) succeed and will provide both a core vision and purpose.

- A poor leader will always focus on authority.

- Do you lead by example? Good leaders will be respected for their high ethical values, passion and endeavour and their success will inspire others to follow.

- A poor leader will constantly fail to deliver positive results and will be unable to inspire others.

- Are you humble (without ego)? Good leaders, throughout history, whether on the battlefield, in politics, or in commerce and industry, have always put service before self. Sometimes, this is to their own detriment and sometimes at the cost of their own life (see chapter 8 -

Lieutenant Colonel Herbert (H) Jones). It should always be about the team and the best way of accomplishing the task(s), by bringing individual team members along together. A good leader will, when needed, just roll up their sleeves, get their hands dirty and work with the team to achieve the best possible result.

A poor leader will often act as if she/he is the only one who can perform the task(s) correctly and micromanage everything. They normally end up doing it themselves rather than trusting others to do their best and getting it done their way. One of the worst things about a poor leader is the use of their position or status to control others, particularly when they are arrogant and/or let their ego or pride get in the way of the task.

- Do you have empathy? Good leaders show both empathy and respect, being prepared to step into the other person's shoes and valuing different perspectives and points of view.

A poor leader always seems to show disrespect for their team, blinded by their own ego, and invariably the team members end up despising them.

- Can you communicate? Good leaders will always understand, instinctively, that effective communication is a two-way street and requires openness and the ability to listen. They understand that it is not just about getting their own point across. It is also about open dialogue, with the single caveat that, in the heat of battle, it is about following orders.

A poor leader is a poor communicator. It is, unfortunately, all too common for closed-minded leaders, who are usually also unable to accept criticism, to close their ears to the views of others.

- Are you flexible? Good leaders understand and deliver a flexible approach and a contextual management style.

A poor leader will treat team members as a commodity or pit them against each other. They tend to approach issues on the basis that it's 'their way or the highway', which tends to generate a feeling of dissatisfaction and low morale. Obviously, there will be times when the rules must be followed but sometimes those rules might be inappropriate and need adapting or even changing.

Repeated mistakes

We just do not learn, do we?

The major incidents and disasters that I have discussed so far, and those to come, contain repeated mistakes and that just makes me want to cry with total frustration. They all seem to start from those near miss events I spoke about earlier. You know, the ones that the team leaders, at whatever management level in an organisation, should be observing and then controlling, well before they actually hurt someone or something.

A very good friend of mine, Doug Holman, noticed a similar phenomenon in the consistently low business survival rates. He learned from experts in finance and insolvency that a relatively small number of well-understood and avoidable mistakes were not being addressed, either by the business community or the government departments gathering the data. Never mind the bankruptcies, redundancies, lost homes, family breakdowns and unpaid creditors. And let us just ignore the underperformance caused by lack of confidence and extreme risk aversion. If the number of startups exceeded the number of closures, why worry?

With a background in aviation safety, including spells with manufacturers of flight data recorders and other 'black boxes', he saw an opportunity to share the same technique, which has reduced the number of aircraft accidents and incidents, with the wider business community – the use of simulators. Along the way, over a bottle of wine (well, it might have been a few actually), he and I discussed how the integration of risk and opportunity were totally intertwined into all corporate activity. We soon realised that the assessment and management of all forms of risk was key to business success. This included sales, financial, operational, people and reputational risk. These relate directly to what is otherwise known as the balanced business scorecard. This is a tool for monitoring corporate strategic decisions based on aspects relating to financial, customer, internal processes, and learning and growth indicators.

We decided to field trial an early iteration of the software, using one of my construction sector clients that was looking to roll out an incident investigation course throughout their organisation. The company was very happy to engage with us and although embryonic at that time, attracted a great deal of acclaim from all those taking part.

We cross-sliced the organisation and the course delegates were drawn from all departments. We formed the delegates into teams of five, with one person from each department (ie sales, accounts, operations and HR) represented, plus one director. We then set exercises based on several pre-set parameters, some fixed and some flexible.

Each session (about 60 minutes) represented one year of corporate results and teams were required to compete. The winning team would be the one with the best results, overall, at the end of the full five-year business cycle. What the delegates were not aware of was that Doug and I were intervening at the end of each session. We left them alone after the year one session, just so they got the idea of the game and to develop the competitive spirit. At the end of the year two session, we introduced a major corporate disaster, and at the end of year three, an enforcement authority prosecution. The fixed parameters were fixed for a reason, to simulate reality, as far as we could foresee, based on a rough and ready PESTLE (Political, Economic, Sociological, Technological, Legal and Environmental) analysis.

The teams were then required to manipulate the remaining business parameters, based on the balanced scorecard. They needed to both recover from the incident and ensure the ongoing survival of the company. Bearing in mind each team was in competition, once the annual scores came in it was clear to see a massive divergence in year three. However, as each session was debriefed, we could also see how the most successful team had 'played' with their parameters and how the other teams had learned from their own mistakes. Over the remaining two years, the results started to converge and, collectively, all of the teams managed to achieve the target of both recovery and sustainability.

The risk management variant of The Business Game™ was a great hit and once the rest of the workforce knew that prizes (a bottle of nice wine for each member of the winning team) were being provided, the word spread and everyone wanted to play. The result was that the company changed its whole safety management system to integrate and embed the learning into its standard operating procedures. Unfortunately for me and Doug, the company went from strength to strength and was eventually acquired. As with all mergers and acquisitions, the larger company imposed their traditional methods and considered that playing games was not a good use of staff time. Although my original client company is still in existence, it is not the dynamic organisation it was. Internal training is now undertaken in a much more traditional way, just to tick the compliance box. Our feedback from the training manager is that the original players are now bored with the 'same old, same old' methods.

Doug and I continue to develop the risk management applications and hopefully these will have been released into the marketplace by the time you read this book.

CHAPTER 5

CARE IS NOT JUST A DUTY

The 'duty' of care is probably not the best standard

You will probably be scratching your head at this point and thinking, why would he make such a ridiculous statement? Everyone understands the concept of a duty of care. You would, of course, be right, in principle at least.

However, as I have already demonstrated, history has shown that those with responsibility over others have repeatedly failed. It does not matter if the responsibility relates to parental care, or the direct obligation contained in a contract of employment, or implied obligations to our neighbours, these 'duty' holders have simply failed. Like me, you will be able to cite examples of workers being treated like slaves, as well as callous individuals who are just too preoccupied with their own wellbeing to care. This could even include parents abusing their own children. These types of human failings normally hit the headlines but that is not my point.

In response to these failures, the courts have, quite rightly, imposed a 'duty' of care on the care givers, to ensure that they provide the level of care required, when considered by that 'man on the Clapham omnibus'. I am sure you remember him (chapter 1) and this is the 'duty' of care that I am asking you to consider. What do you think might happen when someone who should care, morally, does not care and a 'duty' is imposed on them?

Firstly, those people who are, thankfully, in a small minority, will simply try to avoid it at any cost. It does not matter to them that they might be found out and caught, let alone punished. Then, there is a much larger group of people who see care as simply a compliance thing. They will produce a form, usually based on a tick box, to either jog their memory or, worse, to provide evidence to show that they have exercised their 'duty' of care just in case something goes wrong.

Accordingly, my response to the perfectly valid opening statement is that those who do care, as part of their natural human response mechanisms, do not need to be forced into it by the imposition of a 'duty'. The 'duty' is only there to ensure that those who are owed a duty are provided with a legal fall-back position, and in the hope of persuading those who might not have considered it to think about it, at least. However, there is a big 'but' here, as the 'duty' has never ensured and will never ensure that the care needed is actually given, or abuse is actually prevented. Those inclined to wilfully neglect or evade the 'duty' will not be coerced into doing so, and accordingly, there needs to be a legal 'duty' so that the victim has a process to obtain some sort of remedy.

It is a very straightforward process.

Firstly, it must be decided if a 'duty' was owed at all, and although a simple matter, where a contract of employment or a contract for service exists, it can be more complicated in other cases. Secondly, it must be decided if the 'duty' was breached, and again, a simple matter on the face of it, where contracts exist but much case law on the matter has been developed over hundreds of years. Thirdly and lastly, harm needs to have occurred and in a civil case, once all three matters have been decided in favour of the victim and on the 'balance of probabilities', compensation is payable.

The legal duty goes a step further as well, by requiring employers to ensure that the actions or inactions of each one of their workers, in the course of their work, does not cause harm to anyone affected by their workplace activities. Accordingly, insurance is often a minimum requirement for doing business in many countries, to ensure that any organisation acting negligently can pay compensation to the victims. Not to cause confusion but you should also be aware that several defences are available to a claim of negligence and will be investigated by the insurers when they feel they can rebut a claim.

A total defence of 'volenti non fit injuria' exists, reflecting a common sense notion that 'to one who is willing, no harm is done' and requires the respondent to prove that the victim was responsible for his or her own harm. There is also a partial defence available, being 'contributory negligence', which aims to reduce the damages the respondent must pay and suggests that individuals owe themselves a duty of care as well. The defence of illegality denies recovery to certain claimants on the grounds that their claim is tainted by their own illegal conduct. and one party may seek to exclude all potential liability to another party in advance, but take care, as some contractual exclusions might prove to be unfair and/or unlawful themselves.

So, imposing a 'duty' of care is not that straightforward after all.

State intervention

Most countries around the world have imposed a 'statutory duty' of care, by enacted various pieces of enforcement legislation. In the case of allegations of a criminal breach of the 'statutory duty', an even higher standard of 'beyond reasonable doubt' is required, as the remedy here is punishment, such as fines and/or imprisonment. It is worth noting that it is impossible to obtain insurance for any 'criminal' breach of the 'statutory duty' of care.

In the UK, the relevant primary legislation is the Health and Safety at Work etc Act 1974, the Environmental Protection Act 1990 and the Regulatory Reform (Fire Safety) Order 2005, together with many pieces of associated primary and secondary legislation, supporting codes of practice and guidance documents. Much of this derives from the case law and/or subsequent serious incidents, and prior to 31 December 2020, several European Union (EU) regulations and directives. The enforcement of these 'statutory duties' is undertaken by a range of enforcement authorities and although the approach to compliance, in the UK at least, is functional (ie a 'reasonable' assessment of the degree of harm and controls), as opposed to being prescriptive, there are still some absolute requirements and minimum standards that need to be achieved.

In many other countries of late, the message from the state is that organisations and individuals will be named, shamed and punished for breaches of the various 'statutory duties' and these punishments are becoming ever more severe. The general approach now is that the punishment needs to be based on perceived risk or potential harm (ie realisation of harm is no longer an absolute prerequisite) and the ability to pay, which for some organisations translates into millions of pounds (sterling). Sometimes, in the most serious cases, the organisation is wound up and the individuals ultimately responsible are given a custodial sentence.

The message now is very clear and very stark. If you do not care as you should, you will pay the cost of the harm, even if that harm has not yet occurred.

Parents do not need a 'duty' to care

Every now and again, we hear of a parent that fails to care or, worse, abuses a child, but thankfully these instances are very rare indeed. They cause society so much anguish when they do occur. So, if we expect this standard to be achieved without state intervention as the norm, why can't employers just do it by way of a moral obligation in the workplace?

The analogy of comparing family life with the work life is also a very useful one, as we generally spend as much, if not more, time at work as we do at home. Furthermore, our work colleagues routinely knit together as a team, becoming just like a family, with similar trials and tribulations. Not totally dissimilar is the love that is expected within the family setting and, although very seldom expressed that way, the team should feel love, or at the very least a loving team spirit towards each other as well.

Take, for instance, the parental response when a child is late home from school. Let us just work through, in our heads, the sort of human reactions as the situation develops. At 4.00pm say, Jack or Jill or the name of your son, daughter, brother or sister, is expected home but they fail to show. It is quite likely that you will think to yourself, I wonder what has happened, and maybe you will think, oh well, late again, playing with their friends or side-tracked somehow. Not long now. It gets to 4.15pm and you think just wait until she/ he gets in, you are going to give him or her a piece of your mind, especially as it is winter and getting dark. Come 4.30pm and you are now getting a bit edgy and think, perhaps I should call, as all kids have mobiles nowadays, don't they?

Maybe you would have done it sooner. However, at some stage you make that call and there is no answer. Now the alarm bell starts to ring – not the tuning fork this time – and you will not ignore this warning. OK, you think calm down and act logically. What you were doing is put on hold and now furthest from your mind, so you start by calling the siblings, if there are any, or known friends. Either no response or a negative response. The alarm bell starts to get even louder and you're feeling sick. It is now 4.45pm. Think. Where could she/he be? Next point of call are the friends' houses and other parents. No, nothing. She/he must have an after-school activity or has received a detention, but hold on, the school always calls if something like that happens and normally lets you know in advance. Still, you decide to give it a try anyway, but no, they have no idea where they might be either. I am pretty sure that most parents will be panicking by now as it is fast approaching an hour, so your last-ditch effort before calling the police is to call your partner or next of kin.

Thank heaven, they answer. However, to your horror, they are both at McDonald's having a burger. Even though you are totally relieved, the adrenaline starts to pump again but this time it's the fight response and anger takes over. You are going to give them both hell when they get home. By the time they walk through the door though, the adrenaline has subsided to be replaced by a headache, but your worst fears did not materialise. You are so pleased to have them home that you just burst into tears and hug them both so hard.

Did you need a 'duty' to do that?

No, of course not. So, why does that same story not play out in the workplace when, for instance, a lone worker does not check in or respond to a call at the end of the day?

The International Labour Organisation (ILO) estimates that 2.3 million work-related fatalities happen each year, with about 15% of them related to lone working.

In April 2017, a UK water company was fined £1.8 million following the death of a lone worker who drowned in a filtration tank. The father of two children died in December 2013, when he was unblocking a filter at a water treatment works. He fell through a hole into just over six feet (2m) of water. The irony was that he activated his lone worker alarm system but it was not until he failed to respond to a call, some 90 minutes later, that someone was sent to investigate. They were the ones who found him dead. The judge was also told that the water company had been made aware of the dangers of falling into tanks on several occasions.

Teaching the family and the team to care

While discussing analogies between family life and work life, how about considering how you might teach your children to cross the road?

Again, you will probably not require a 'duty' to do this, as it comes naturally. Even if you have no children of your own, I am sure you were one once, so just cast your mind back to when you first ventured out on foot with one or both of your parents or guardians. It was probably with both of your hands being held by each of them, if there were two, or on a set of reins to make sure you were totally protected. As soon as you could understand, you were probably told about the dangers of crossing the road and the 'green cross code', as it was known in my time, or similar. You would have been told to find a safe place to cross first. Then you would have been told to use all your senses when looking right, left and then right again, bearing in mind that I was taught this in England where we drive on the left and it might be the other way round for you.

If you saw or heard something coming, you were told to stop and wait for a safe gap in the traffic, with a reminder that bicycles and the like, particularly now electric cars, do not make much noise. Once you started to cross, you were told to keep looking and listening until you reached the other side. Eventually, this process would have become second nature, through constant reinforcement and practice, well before you were let out on your own.

I would like you to remember this analogy, as we will be visiting it again when we discuss worker competence (see chapter 7). For the moment though, there is just one piece missing and that is the need for discipline. I am sure you will also recall that when you forgot the 'green cross code', or whatever it was called for you, one of your parents or guardians would shout "stop" and, immediately after you had been prevented from walking in front of a vehicle, the message would be reinforced.

Hopefully, you will see that the love of a parent or guardian is, or should be, like the love of a caring employer. We spoke about good leaders in chapter 3 and those same traits are virtually identical to those of good parents.

Discipline and care

With just over 30 years as a serving magistrate (I am still a Justice of the Peace, albeit now on the reserve list), spending a good proportion of that time in the Youth Court, I was constantly amazed at how some parents stood by while their sons and daughters went off the rails. With a few notable exceptions, they were all very loving but just did not know where or when to set boundaries. The problem is that children are always testing those boundaries. I am not advocating any form of violence, but the welfare of the child, in both the legal and moral sense, requires that parents need to impose discipline when appropriate boundaries have been crossed.

The same applies in the workplace and a significant minority of workers will act just like children, by testing boundaries. If the workplace rules (see chapter 9) are set appropriately, then it is incumbent on you to ensure that they are followed, with appropriate sanctions for those who break them, particularly if they are broken repeatedly. Sometimes, all that is required is a gentle tap on the shoulder, or a nod, or even what I call a 'fireside chat', to inform the transgressor of the error of their ways. At the other end of the scale is punishment and, again, I am not advocating violence. However, for a punishment to be effective, it must always hurt, figuratively speaking. The ultimate punishment (no, I am not talking about the death penalty) is incarceration and for some breaches of the rules, such as a statutory offence (crime), this could mean a prison sentence, even if it is suspended.

In both the family and workplace setting though, I am more concerned about the lack of internal discipline or turning a blind eye to rule breaking. Sometimes, worse, is the inconsistency in administering consistent discipline. Failure to do this can, and normally does, lead to parents or workplace supervisors being played off against each other.

Back in my day job, I was called upon on one occasion to advise the defence team at a criminal prosecution, where my client attempted to defend his position by claiming that the reason a worker received serious injuries was that he refused to wear the personal protective equipment (PPE) that had been provided. It was stated, in open court, that the injured person was "told by the supervisor, until he was blue in the face, to wear his PPE" and the person continued to ignore the instruction. The judge turned to my client and asked, very simply, "Why did you allow the tail to wag the dog?" as it was the responsibility of the employer to ensure the safety of the workers. The operative part of the sentence being 'ensure', which put simply means to make it happen.

I have had cause to mention this story to several of my clients and the analogy becomes:

- Tell someone once and the responsibility shifts to them to carry out the instruction

- Tell someone twice and the responsibility is shared 50%

- Tell someone 10 times and the responsibility will revert to the person doing the telling 90% percent and the person being told 10%

- Tell someone 'until you are blue in the face' and the responsibility is all yours

This analogy holds true in both the workplace and family environments, so if you are prone to repeating yourself, it is worth bearing this in mind and stop wasting your breath. It is doing more harm than good.

The downside when a 'duty' is imposed

I believe that a significant minority of workers, at least, undertake their workplace activities in the oblivious state of knowing that their employers owe them a duty of care. In so doing, they fail to grasp the idea that they owe the same duty to themselves. Even when that tuning fork vibrates, instead of going into freeze mode, they just carry on regardless, overriding that funny feeling inside. They just go about their task(s) believing that someone else is responsible for their safety and will look after them.

It just does not work that way.

Figuratively speaking, the good leaders teach the team members to fish and do not do the fishing for them. The poor leaders rarely even bother to teach the team and just let them find out how to fish for themselves. In both cases, the team are all concentrating on doing their own fishing and when one of the team catches a shark with great big teeth, they are on their own. Well, they are in the initial stages. You might well ask, where is the supervision?

Even those charged with supervision cannot be everywhere. Unless a worker who is not yet competent, perhaps a new starter, apprentice or young person, is under close supervision, most workers are, or should be, competent enough to undertake their own work (fishing) unsupervised. As you will see in chapter 7, this is, by definition, an important element of being competent. Remember the analogy of taking a child out for their first walk on the pavement and crossing the road. Until they are competent, they are on a lead or having their hand held. At some stage, when the time is right, they need to be left to their own devices.

Unfortunately, even when they are competent and doing their own thing correctly, others can do silly or dangerous things around them. Someone could be driving too fast and mount the pavement. Even the competent person, in those instances, can get caught out and the same goes for the workplace. Take, for instance, that incident where the operator got trapped in the car transporter. It is very likely that one of the team was the team leader and they both decided to break the rules. Each weekend, one of them would tick the box that said the two-person operation was in place. It was only when the inevitable happened that the one who ticked the box died.

So, even when a 'duty' is imposed, some members of the team, hopefully not you, will do their level best to circumnavigate the safe operating procedure. Sometimes, it could even be to help the team achieve the task(s). Sometimes, the SOP does not reflect reality. Sometimes, it is because someone is too lazy to do things properly and sometimes it can be an intent to deceive. How often have you seen a maintenance sheet with the same statement, something like 'All OK' or 'N/A' or 'Ditto', repeated, line after line and page after page?

Some people believe that their duty is fulfilled when they enter that tick in the box. However, I wonder if they ever ask why nothing is ever discovered that could suggest that all is 'not' OK. I wonder if they ever double-check that the inspection has been done properly. Worse still, can anyone be certain that the inspections are being done at all?

As you will see, I hope, imposing a duty on someone does not always mean that the organisation or the individual cares at all.

Self-preservation

However, this is all about you. So, just consider this for a moment. Have you have ever entered a room or space and thought to yourself, hmm, there's something not quite right here?

You do not yet know what it is but you do know that something is not right. Well, you also know, as we have already discussed, that your tuning fork is telling you to freeze. Our ancestors would listen to it and act on it. They would respond to the adrenaline rush by sharpening their senses in readiness for the impending fight or flight mode. What did you do?

So many of your colleagues will probably do the same. They will just ignore it, believing that it is the 'duty' of someone else to look after them. The 'duty' of care rests with the boss, or the building owner, or someone else. Anyone else. They stop looking after themselves and simply carry on, oblivious to what their own body is telling them.

During my career as a firefighter and later as a senior fire officer, I had cause to rely on my tuning fork so many times. One incident I recall occurred at a fire in a tyre warehouse in Deptford, south London. I was part of a specialist crew that was deployed to enter the basement in extended duration breathing apparatus. We were tasked with getting to the bottom of a deep-seated fire.

As you might be able to imagine, burning tyres produce thick black oily smoke and it is impossible to see anything at all. We progressed slowly, in the darkness and searing heat, hauling a line of charged hose. This can be exhausting under normal circumstances, let alone in these conditions. The building was old and like a rabbit warren, with lots of short passageways and stairs that turned left, right and back again. After what seemed like hours, but in reality it was only about 10-15 minutes, something changed. To this day, I do not know what made me do something that went against all my training and experience. I removed my face mask.

This was a tyre warehouse, remember, and our face masks were totally obliterated by black tyre residue so we could see nothing. To my amazement, we had missed the seat of the fire completely and there was no smoke at all in the area where we were all walking. It was not a walk as such but what is known as a firefighter shuffle. We were all attached to each other with our personal lifelines and taking our weight on our back foot so as the front foot could feel the way. It must have looked quite a sight to the onlookers, as we were all now shuffling across the main shop window.

I called out to the other team members, who were all still oblivious to what I knew, and with great reluctance they all joined me in removing their face masks. We all stood there, transfixed on what we saw on the other side of the shop window. When we looked out at the crowd that had developed, they all looked back at us and started clapping. From their perspective, it must have been hilarious to watch, but for us it was rather embarrassing. I will never forget it.

Toilet paper

Other experiences have proved to me that circumstances do not always turn out this well or this funny. The tuning fork works and we are all lucky enough to possess one. It is there for a reason and you would do well to use it, or should I say listen and react to it. If you choose not to, you will lose it, to your detriment.

Once you do start to pay attention, you will find that less will go wrong, even though there are, sometimes, unforeseen events. There will, of course, be those around you who will get in the way. They are probably not listening to their own tuning fork. They might also be going about their business believing that you will be doing their looking, feeling and thinking for them. There are also those who believe, wrongly, that the policies and procedures will ensure, on their own, that all is well.

In most cases that I have encountered as a safety advisor to numerous clients, most members of staff do not have a clue what has been documented. In many instances, they are not even aware that these documents even exist. If they do know they exist, most will not have read them, even if they are required to sign a document to confirm that they have read and understood them. Very few workers would ever refuse to sign such a document, especially if they have only just started working for an organisation.

Furthermore, a piece of paper, or e-document more like, has never stopped an unsafe act or omission, or dangerous condition. Ironically, it can be more advantageous in these circumstances to plead ignorance of the 'statutory duty' to produce one than to present such a document after an incident. It suggests that you knew what you should or should not have done but went ahead anyway. As the well-known adage goes, "Ignorance of the law is no excuse." However, it does provide some damned fine mitigation.

The key to caring is implementation and without it you are lost. The 'tick box' culture of producing documents and failing to carry out what they require is proof that you simply do not care.

Think of the tick box paperwork like toilet paper. Imagine that you are sitting on the toilet, just contemplating your navel, or something on the wall, and at the very moment you reach out for the toilet paper, you find that it is missing. Oh no, what are you going to do? You look for the spare roll but no, none around. You have no choice but to improvise and yes, I can imagine what you are thinking at this stage. The point is that the toilet paper is there for a simple reason and I will let you work that one out for yourself, without prompting.

It is when you cannot perform that primary activity, or when things go wrong, that the toilet paper proves its worth. Again, I will leave you to work that one out. The toilet paper is an important commodity and you are lost when it is not there. It is there to clear up after you once you have performed the task in hand (sorry). So, the tick box approach to the 'duty' of care is just like the toilet paper. Without it you will need to improvise, If the paperwork is not provided, replaced when necessary and used properly, you will not be able to clean up afterwards.

Cleaning up afterwards is simply about proving that you cared in the first place so, surely, isn't it just so much easier to care?

Inappropriate and inaccurate information

I do realise that you are hoping that I have finished with the toilet paper theme. Sorry, I am going to disappoint you for a little bit longer, as it works quite well in explaining my point.

Complying with the 'duty' of care, as opposed to simply caring, invariably means that you will need to provide the necessary evidence (toilet paper) to prove that you have complied with that 'duty'. We have already discussed the problem with operatives and supervisors simply ticking boxes but you must prove, to your bosses, that your key performance indicator (KPI) targets have been met.

I do sincerely hope that your health and safety KPIs are not related to your promotion or salary, as I fear that you might not be as truthful as you would like to believe you are. You may well be under pressure from middle managers, or senior managers if you are a middle manager, to impress or at least not disappoint. You might even want to altruistically shine a good light on your department and your team. Only you know, in these circumstances, if the KPI targets were appropriate, useful and met.

However, let me now suggest one KPI that is fraught with danger. The 'zero accident' target. I can virtually guarantee that this is the most unhelpful and misleading target in the world, for a very simple reason. With the single exception of a fatality that, in most civilised countries, is particularly difficult to hide, supervisors and managers under pressure to show good KPI results regularly and routinely ensure that accidents are simply not reported.

I have even experienced a situation where, following a major injury, a safety noticeboard at the main entrance of a large utility site was modified to show a continuing 'zero accident' record from the day after the incident. It was also widely known on the site, along the lines of the research undertaken by Heinrich, Bird and others, as discussed previously, that there were regular near miss (hit) incidents which went unreported. In fact, any reporting of unsafe acts and/or unsafe conditions was frowned on, for fear of the amount of toilet paper it produced and the downtime involved. Worse though was the high number of minor injuries and anecdotal knowledge of the unspoken work-related ill-health. The re-dating of the noticeboard also kept from public gaze, at least, the frequency of major incidents and even a double fatality, albeit some years previously.

In the words attributed to Mark Twain (falsely, it is understood), "there are lies, damned lies and statistics", which is what was going on at this site, as I suspect is also happening elsewhere. It is much better, in my view, to record everything and learn from the findings rather than cover it up and learn nothing until it is too late. In fact, a good rule of thumb is to produce 'as little as possible and as much as is necessary'. As little as possible speaks for itself but how much is necessary?

To determine how detailed the evidence needs to be depends on its ultimate purpose. So, if it is required to indicate what types of near miss incidents are occurring, it is reasonable to simply collect some very basic information, for compilation and statistical analysis. If someone has been harmed, then it depends on the severity of that harm. For instance, a cut finger warrants a simple recording of the event, just like the near miss. If the harm is more serious, a more thorough investigation and root cause analysis will probably be required, to ensure that a reoccurrence of the incident does not occur. As the seriousness of the incident escalates, other interested parties, such as senior managers or the organisation's insurers, will need to see the report, so it will need to be more informative. If the likely reader is the enforcement authority, it will need to be very detailed. If the ultimate arbiter is likely to be a coroner or a judge, you will need the full chapter and verse.

If you are writing the types of report mentioned in the latter cases, you are also the one that is likely to be called on to explain to the person's family why their loved one will not be returning home as expected. Let us both hope that you never have to do that as I can tell you, from bitter experience, it is one of the worst jobs in the world. They will also be reading your report(s), line by line, when they recover.

CHAPTER 6

BAD RULES ARE NOT GOOD

What bad rules look like

Sometimes, a bad rule is easy to spot, as it is so totally ridiculous. It just beggars belief why it should ever have become a rule in the first place.

For instance, one of my clients is contracted to fit carpets in large complexes. I received a call asking if the health and safety (H&S) legislation required the fitters to wear a safety helmet and safety boots. My response was that, although it ultimately depended on an assessment of the risks involved and the likelihood of harm, it seemed doubtful, bearing in mind the nature of the activity. He asked me to discuss the situation with the principal contractor (PC), who shall remain nameless, which I did, as that is what he was being told by their safety team. The problem was that the building had only recently been built and although fully fitted out, apart from the carpets, not yet fully handed over to the owner.

Quite understandably, during the construction phase, the usual 'No Hard Hat, No Boots, No Work' rule applied, which was strictly enforced by the building owner's health and safety manager (HSM). So far so good. The problem was that although the construction work, as it would be understood by most people, had been fully completed, including all snagging (the final bits and pieces), the HSM insisted that it remained a construction site and the rule still applied. I tried my very best to reason with him on the basis of there being no risk whatsoever to the carpet fitters, but he was adamant and reverted to the contract.

I had no choice but to explain to my client that the rule being adopted was, at this stage, nothing to do with safety but a contractual obligation. In effect, if the contract stated that they would only be allowed to lay carpet while wearing a tricorn hat, pink frilly dress and clogs, then that is what they were obligated to do. The fact that the operatives found it virtually impossible to lay the carpet while wearing a hard hat and steel toe-capped boots was immaterial. It was completely stupid but my client had no choice. So, he simply broke the rule. He stationed a lookout, to ensure that the fitters were wearing their PPE, ostensibly while at a tea break, whenever the HSM was around.

Inappropriate and unlawful rules

Sometimes, a rule is just inappropriate for the circumstances. Unlike the one just mentioned, which was both plain stupid as well as being inappropriate, others are harder to identify.

For instance, a rule that was appropriate may well have been changed following a change of circumstances but has not been implemented locally. It may be that someone overseeing the plain stupid scenario of the carpet fitting saw the light of day and changed the rule. Unfortunately, the HSM went to another site and implemented the original rule, which has now been changed. This is called issue control and the rule being enforced is both out of date and inappropriate. Unfortunately, in this instance it will remain stupid.

Issue control is a serious problem though, especially where H&S rules are concerned, as some people follow rules to the letter without thinking and nothing triggers the tuning fork, as it all looks perfectly reasonable. The issue is that someone is now following flawed rules. Even when the tuning fork does start vibrating, most people will not be able to see clearly what might be wrong. Similarly, some rules are misunderstood and enforced incorrectly. This can be done where the rule maker does not understand why they are implementing a rule other than 'the law' says she/he must.

Take the situation with manual handling, where one of the statutory requirements in the UK is for workers not to lift anything beyond their capability. Guidance has been issued that suggests what weights can be handled safely by most people. However, firstly, these are gender specific, as most men are capable of handling larger loads than most women, although it does not suggest that this is the only scenario. Secondly, these guideline weights are just that, guidelines, and apply only in normal conditions. So, when rules are developed that state maximum loads that can be lifted, this load may be well within the capacity of a female employee but beyond that of a male worker.

This prescriptive rule may well be inappropriate in the individual circumstances and potentially unlawful as well, by way of sexual discrimination. The common sense approach would be, therefore, to assess the individuals concerned wherever loads require moving beyond the guidelines provided. You should not forget, either, that individual circumstances might also change, such as the member of staff becomes pregnant or feels poorly and may well need to cease all handling operations at that time.

There are many situations where inappropriate or unlawful rules exist and you should be ever vigilant to make sure that you are not caught out, or even condone the bad rule. If you know the rule is bad, get it changed.

Procedures are rules

Rules do not always appear on signs and notices. Sometimes, they also exist as unwritten rules and we have already discussed the problem with custom and practice and 'grandfather rights'.

How many times have you been confronted with so many procedures that you just cannot possibly take them all in?

It is just so sad that so many people try to cover off all eventualities with toilet paper (oops, you probably thought I would not go there again). I am sure that you have also seen the outcome of plagiarising the rules and procedures of others, sometimes also including a straight copy and paste of the legislation, where the result is a tome that will never be read, let alone understood.

The problem is that the human brain is incapable of handling much more than about nine things at once, and where multiple rules are concerned, it is accepted wisdom that a good set of rules would constitute no more than five dos and five don'ts. Anything else will just get filtered out or forgotten, very quickly. The art of producing good rules is to keep them short and simple. Just like the general rule of thumb for all documentation – as little as possible and as much as is necessary. Under normal circumstances, the simple rule 'No Hard Hat, No Boots, No Work' is a good rule when correctly applied.

Sometimes, the document or rule writer falls into the trap of saying too much. How many times have you seen the policies and procedures looking like a straight take from the legislation, or worse, attempting to cover every single eventuality?

The problem here is that a, thankfully, small number of members of staff might take the view that if a task is not fully addressed, or addressed at all, it is a management failure, with the potential to make a civil claim, as a breach of the employer's 'duty' of care. In fact, only 'significant' risks need to be addressed. This became custom and practice in the fire service and far too many claims were paid out of court when firefighters twisted their ankles and fingers while playing volleyball in their stand-down periods.

This practice was prevalent back in the 1980s and what can only be termed abuse of the system caused the principal managers to call a stop. A decision was taken to allow an arbitrary case to proceed all the way to an employment tribunal, where the tribunal chairman stated that no formal procedures could be developed that accounted for every possible variation of a likely event. The tribunal panel went further and suggested that all firefighters required basic physical fitness training, and shortly thereafter, one firefighter on each of the four watches on a fire station received specialist training to become a physical training instructor (PTI).

The role of the PTI was to ensure that all physical training, including volleyball, was delivered in a consistent manner and logged in a personal record book. There were no detailed procedures and as soon as the firefighters became aware of the resistance by the principal managers to any more spurious personal injury (PI) claims, the practice stopped.

The moral of this story is, therefore, that 'less is more' and very concise, task-relevant rules and procedures are much better than long-winded tomes of information that can be picked over by those with an axe to grind. Keep the wording short, say only what needs to be said and address everything else by ensuring that all members of staff are competent for the role (see chapter 7).

Some procedures get in the way

I firmly believe that blindly following procedures can be dangerous.

One of my consultancy contracts involved working with a 'Lean' expert, using a methodology to secure efficiency savings and effectiveness improvements, which focuses on reducing and eliminating waste, as part of a transformation exercise for a hospital Trust. Working collaboratively with the other consultant, it was decided to set up a cross-sectional group of medical and technical practitioners. The first thing we discovered, on day one of the programme, was that the participants were unaware of the true nature of what each other did and in some cases, an acrimonious relationship existed between them and their departments (silos).

Once we resolved the personality issues, we asked each departmental representative for an example of where they felt they could improve efficiency and/or effectiveness. A standard consultancy approach I know, but it is one that normally gets straight to the obvious low-hanging fruit and provides some quick wins. There was one story that stuck out. It was delivered by a locum surgeon, under cover of the Chatham House rule, that those in the room were "free to use the information received but neither the identity nor the affiliation of the speaker(s), nor that of any other participant, may be revealed". Accordingly, not that I would anyway, but I share this story without divulging the identity of any of the participants or the Trust involved.

So, bearing in mind that this was happening in the National Health Service (NHS), my colleague and I, together with most of the others in the room, assumed that the medical procedures in all operating theatres would have been very well tried and tested. At the time when the information was being shared, it was understood that certain anomalies were already being addressed, including the one I am about to share and I am hopeful (a pity I cannot say confident) that was the case.

Apparently, it was then standard practice in the north of England, during an operation to amputate a single limb, to mark up the good one as a warning not to touch it. Whereas, yes, you are well ahead of me, in the south of England it was standard practice to mark up the bad limb for removal. As you can imagine, the room fell silent. Initially, we all thought it was a joke and laughed, hesitantly, in absolute shock and horror, but no, it most certainly was not a joke and no laughing matter. Luckily, the clash of procedures was, so we were told, already being addressed at the highest level but it begs the question: How many poor souls were on the receiving end of this malpractice?

I am sure we could find out but that is not the point of the story. The point is to show that blindly following procedures is dangerous. For those poor souls subjected to this lack of common sense, they were left without their one good limb. During the rest of the transformation programme, you will be pleased to note, I am sure, that no more diabolical tales of this sort arose. Much was learned about the bureaucratic processes that get in the way though. In fact, on the back of this sad story, we established the principle of "efficiency and effectiveness, in a safe environment", which has also now become one of my company's straplines.

The concept of 'Lean' process improvement was originally developed by Toyota, to improve the operation of their production processes by cutting the time it took from receiving an order to its ultimate delivery. It is a continuous system, with a focus on reducing and eliminating waste. However, we discovered very quickly that some waste in the system was essential. In many respects, many organisations perceive safety as an irritation that just gets in the way of efficiency and effectiveness.

Try telling that to the poor patient without any remaining legs. The removal of the patient's good leg was undertaken very efficiently and effectively. It was not a safe practice though and cost both the patient and ultimately the hospital dear. So, instead of a focus on just waste, we agreed on the concept that waste could either add value or not. Safety, when reasonably applied, is not waste. Wrapping an organisation in cotton wool is waste and adds no value. What we found was the need to eliminate non-value-added waste, like bureaucracy for the sake of bureaucracy.

The human response

Human interaction and empathy might also be considered waste in pure engineering terms, but in any so-called caring environment, common sense would demand that it must never be removed.

It seems to me that in this modern world we are controlled by computers and process. Even when we are engaged in that basic human activity of having a chat with someone. Any intervening call on a mobile phone seems to take priority today, even though most have a voice message facility. Once upon a time, not so long ago, it was considered rude to interrupt. Although the caller, in my example, is unaware of the interruption, the receiver is aware. I will always see it as being treated rudely by the person who just had to answer the call at that moment. Obviously, if they were expecting an urgent or important call, that would be perfectly understood, although even then a bit of pre-warning would be the polite approach.

I fail to understand why people are so driven by and subservient to technology. I do not decry the benefits of the massive advances that have been made, especially in the last few generations. I believe that technological advances have been made to improve our lives and support us. Instead, we are forced to listen to inane responses like "the computer says" and that is an end to the matter. How many times have you sought to get technology to do what you want done just to be told that this piece of software or another can only do it a particular way and it is you, the human being, that must adapt?

I have, over the years, sought to develop an IT-based system that would allow my fire assessors to generate their reports in a much more efficient and effective way. They were all used to writing reports but the process was time-consuming. There were so many variables and so many permutations that standard copy and paste methodologies were equally inefficient. Every time I sought the help of a so-called IT expert, all I got was the same old repetitive response. This piece of software can do this part of what you want. Another piece of software can do this part of what you want. There is no software on the market that can do all you want at the same time. Eventually, I had to use a programmer and now, at vast expense, I have a single piece of software that can do what I want it to. No longer do I listen to the computer saying things.

You may well disagree and you would not be the first. It is not too far away when computers will be intelligent and we will all have to listen to what they say. However, I have this old-fashioned notion that the common sense wisdom of the crowd should be considered, at least when it comes to deciding what are acceptable procedures. Process for process' sake is, in my opinion, fraught with danger and, not least, dehumanising. As I say, you may well disagree.

All these little processes are the building blocks of the systems needed to operate and run an organisation efficiently, effectively and safely. In family life, the system that leads to putting food on the table involves many processes, like shopping, preparing, cooking, laying the table and washing up afterwards. The same can be said for business operations, whether private, public or charity sectors, as systems will be required to achieve the core purpose, and one of the most important core values of any organisation should be the safety of its staff and those affected by their activities.

While writing this book, my company developed an occupational safety and health (OSH) management system (OHSMS) for a client, under the auspices of a relatively recent international standard (ISO 45001). Luckily, for me at least, I was proud to have been able to represent the Federation of Small Businesses (FSB) on the British Standards Institute (BSI) committee responsible (HS/1) and understood the fundamental difference between this standard and its predecessor (OHSAS 18001).

Basically, although one of the overriding BSI requirements was alignment with the quality management standard (ISO 9001), the primary focus shifted on to leadership and worker participation. These were the very human elements that topped and tailed the system and without which I would have been loath to engage.

All the way through the development programme with the client, I engaged with the senior management team and worker representatives, to understand what was best for the organisation as opposed to imposing an off-the-shelf product. The client is also a food manufacturer and the OHSMS has also been designed to incorporate environmental management standards (ISO 14001) and food management standards (ISO 22001) as an integrated management system (IMS).

At all stages, the human factors were paramount. The cultural alignment needed to be driven by the human responses, at all levels within the organisation. It was not about process for process' sake.

Never follow inappropriate rules

How many times have you heard the retort, "I was just following the rules"?

Surely, after everything I have said so far, you must be with me by now. You understand what is meant by 'the rules', in terms of both the written and unwritten rules, orders and instructions, signs and pictures. They are either standard or emergency operating procedures, as well as generally accepted customs and "we've always done it that way". You know it can be dangerous to follow an inappropriate or bad rule, either knowingly or unwittingly.

However, did you also know that sometimes following inappropriate or bad rules can be criminal as well. If you can, cast your mind back or refer to the end of the second world war when several officers in the Nazi High Command were put on trial at Nuremberg. Their defence to the war crimes they were accused of was that they were "just following orders" given by a higher authority (the Fuhrer). The judges found that the so-called 'Nuremberg Defence' was, in international military law, under these circumstances no defence. Several of them were executed.

Obviously, the battlefield is not the same as the workplace. There are some similarities though as most people never question rules or orders in the workplace either. They can be routinely open to interpretation though, especially when they are ambiguous or long-winded, and why they keep getting broken. The evolution of rules, in whatever form, is normally driven by the rule breakers. They are based on theories of what was considered best, with the information that was available when they were created. The rule breaker might not have even broken the rule intentionally but it simply did not fit the circumstances at the time. The task still had to get done so a way was found to get it done, in spite of rather than because of the rule.

It is far more important to understand the purpose of a rule than to follow it blindly. This is particularly true if the rule is designed to frame behaviour and ultimately the culture of an organisation. So, the next time you hear a 'terrorist' proclaim, "Why have we got such a ridiculous rule?", you might want to ask him or her why they are making that comment and listen to their answer.

You will also do well to remember that rules are instructions. Together with supporting guidance, they should allow you and your fellow workers to accomplish a task efficiently, effectively and safely. Without this guidance and an appropriate degree of flexibility, the rule could be too restrictive and force you into that tick box response. There may be better ways of achieving that same goal.

You might also like to consider that the person(s) making the rule(s) may not have had the necessary competence. They might even have made the rule to enhance or maintain their own status or an unsafe status quo. If the rule is unwritten, beware, as there is normally a very good reason why it has not been written down.

I am not suggesting that you break all the rules, all the time. I would rather that you understand why that rule is there in the first place. I would recommend that you question the validity and appropriateness of rules, instructions, guidance, signs and routine practices, continuously.

You may well find that you are the one that stands out from the crowd, by flying in the face of the common sense. You might be in a minority of one but someone must do it and that is why you are reading this book.

Be careful though.

Before you exercise your choice to disobey an inappropriate rule, consider how you are going to deal with the consequences. Likewise, before you obey an inappropriate rule, consider how you are going to deal with the consequences.

I never said this was an easy thing to do but it is vital for your safety and that of your family, friends and work colleagues.

CHAPTER 7

RISK, OPPORTUNITY AND INSTINCT

A basic human instinct is to explore, and throughout history we have been taking risks to achieve that goal. From the invention of the wheel and horse-drawn vehicles, to the invention of the internal combustion engine and the car, to the first powered flight and subsequently space travel. These have all started with someone's idea or dream, the instinct to take advantage of the opportunity with a full appreciation of the risks involved.

> *"Twenty years from now you will be more disappointed by the things you didn't do than by the ones you did. So, throw off the bowlines, sail away from the safe harbour, catch the trade winds in your sails. Explore. Dream. Discover."* – Mark Twain

Another American, somewhat later (1959), led many astray when his scriptwriter misread the Chinese language and made a linguistic faux pas, suggesting that risk and opportunity effectively meant the same thing.

> *"The Chinese use two brush strokes to write the word 'crisis.' One brush stroke stands for danger; the other for opportunity. In a crisis, be aware of the danger but recognise the opportunity."*
> – John F Kennedy

However, for the purposes of this chapter, I am going to extrapolate that unfortunate misunderstanding into a loose formula of my own. It is based around the business term concerning a 'return on investment' (ROI), which itself embraces the idea that a reward is available for the opportunity created by the risk. Into that I incorporate the notion of the tuning fork to which I refer throughout the book, which allows me to suggest that:

Risk = Opportunity + Instinct

For me, it provides a more positive perspective and a good thing, as opposed to the generally accepted negative approach to risk as a bad thing. Generally, risk is considered in negative terms as the 'possibility of loss or harm' or 'the realisation of a hazard'. In either case, when risk is understood, it can be dealt with and it is worth noting that it can only be dealt with in four ways. I will explain so that you can decide how you will approach the thousands of opportunities that will present themselves during your life, at home, at play and at work.

You can simply avoid it altogether

There are some risks that are just not worth taking, whatever the reward, even though some still do for reasons only known to themselves. For instance, would common sense not dictate that walking on an unprotected pitched roof without any form of safety equipment should be considered an 'excessive' risk and one not worth taking?

However, free runners do it all the time for fun and I have even seen roofers and tilers doing it, either to save money on the safety equipment or for speed. The reward for these individuals has obviously outweighed the risk, but for most of us we would take the view that the risk far outweighs the reward and conclude that "it's just not worth the risk".

Sometimes though, the activity cannot be avoided. Most of us drive cars, although for my sins I also still ride a fast motorbike. Now, I could sell it and just stick to four wheels or I could leave it in the garage, which is what I do in the winter. That is nothing to do with the increased risk, by the way, but rather, I am just getting too long in the tooth and prefer the comfort of the car during those cold winter months.

In both cases, I have avoided the risk but the opportunity to feel the thrill of the ride is also lost to me. Even in the summer months my instinct tells me that as long as I refrain from riding it at full capacity, on its back wheel, down the high street, all will be fine. I do not do that, by the way. The risk, for me, relates to other motorists and the potential for increased harm should one of them drive into me. Again, for me, that risk is still worth taking and my survival instincts have kept me alive so far. I must admit that I have had a few scary moments along the way but I have learned from every single one of them.

> *"I am always doing that which I cannot do, in order that I may learn how to do it."* – Pablo Picasso

It is impossible to avoid all risk in life and you would probably not want to anyway. Life itself is a risk and you will want to enjoy the experience.

You can find a better way of doing it

More specifically, you can find a safer way of doing it.

If you still feel you want to do something, regardless of how scary it is, or if you are in an industry where the job must be done, regardless, then you will need to consider the safest method of achieving the desired outcome. You might well need to rely on your survival instinct if the task to be performed occurs suddenly. Hopefully, the lessons you have learned already in life will stand you in good stead.

However, most of the time you will be undertaking routine tasks, whether you are a warrior (soldier, sailor or fighter pilot), firefighter, police officer, or member of a lifeboat crew or mountain rescue team. In all walks of life, safe operating procedures are based on risk reduction and a hierarchy of risk controls. This is generally known to all in the safety profession and you will find them in most textbooks, along with the acronym (ERIC-PD), which explains the hierarchy as:

> **E**liminate – As we have already seen, you can just avoid or eliminate the risk altogether but that might not be possible, so you will then need to consider what you must do next. **R**educe or substitute – You can improve the way you do something, do it with something else or do it less often. **I**solate – You could turn off the supply, you could enclose the hazard(s), or keep the individuals likely to be affected away altogether (a bit like eliminate but the risk is still present). **C**ontrol – You should always be considering organisational/ management and technical/engineering controls. These include such things as safe procedures, training and supervision, alongside physical controls like machine guarding, edge protection, local exhaust ventilation, safety devices and specialist tools etc.

Personal **P**rotective **E**quipment (PPE) – As a last resort measure, to cover off those 'residual' risks that ERIC cannot control in any other way, you will need to consider the provision of protective clothing or equipment for every individual that is likely to encounter the primary hazard(s). **D**iscipline – There is no point in providing these safe operating procedures if you do not make sure that all controls are monitored, reviewed and enforced.

The hierarchy of control is exactly that, a hierarchy, which means that you must start at the top of the list and work your way down, choosing those higher up the list in preference.

As an example, imagine a production line where operators need to clean machine rollers regularly using a proprietary chemical to remove the build-up of residue from the processed material(s). You are concerned that the chemical being used has both caustic and toxic properties and an operative might also get his or her finger or hand injured in the rollers.

When considering the hazards, you will find it very easy to get sucked into the provision of PPE as a first resort measure whereas it should always be regarded as a last resort measure. The reason is that, by its very nature, it provides personal protection as opposed to protection for everyone and it does not actually address the nature of the primary hazard(s) at all. So, let us see how (ERIC-PD) might treat the issue.

Eliminate – Why are the rollers being cleaned in the first place? Can the base material supplier improve it to prevent residue? Is there too much pressure on the rollers? Can the rollers be coated with a non-stick material?

The safest control measure is always to eliminate the hazard completely, so if this is possible, there is no need to go any further regarding the cleaning of the rollers, although other hazards might still be present that will also need to be assessed, using (ERIC-PD), of course.

Reduce or substitute – Unfortunately, you establish that you are unable to eliminate the task of cleaning the rollers on this occasion. You must now consider the next item in the hierarchy. Can you make the task safer by doing it less often? Can you reduce the number of operatives that are 'authorised' to clean the rollers? Can you use a less hazardous chemical?

Even if you can do all or just one of these, you will need to keep going, as the hazard is still present.

Isolate – If the rollers are powered then this control measure is essential. Otherwise, can you separate the rollers for cleaning? Can you screen off the machine to isolate the chemical fumes?

Even if this control measure is not possible or practicable, you must still ask yourself the question, as it might just trigger an idea.

Control – Can you implement any technical or organisational controls? These are also known as engineering and managerial controls.

Technical/engineering controls include such things as fixed machine guarding, safety devices (eg lock off arrangements and tools to keep hands clear of the rollers) and local exhaust ventilation (LEV) etc.

Organisational/managerial controls include safe systems of work and standard operating procedures, training and supervision etc.

Personal Protective Equipment – Even if you have managed to find and implement some or all the above control measures, unless you have eliminated the task completely, there may well be some 'residual' hazards (or potential for harm). If that is the case, as a last resort measure, can you provide 'suitable' PPE?

In the case of the roller cleaning, this might include the provision of gloves, eye protection and respirators, but remember, none of these affect the nature of the hazard, they just mitigate any harm in the event of contact.

Discipline – You will now need to make sure that all the above controls are correctly used, maintained, monitored and reviewed. It is also important that all control measures are enforced and that you lead by example, especially when it comes to wearing your PPE.

The above hierarchy should be just a matter of common sense. However, in list form it has always provided a great reminder to me of where to begin and what questions to ask when trying to reduce the risk from any sort of hazard. You might, therefore, feel that it could be a useful acronym to roll out to all your work colleagues and not just your fellow safety practitioners. It will improve the common sense among all the workers.

Competence is more than just 'common' sense

You have seen that the wisdom of the crowd is and can be very powerful in determining what our primary instincts tells us about risk. You have also seen that different societies have different perspectives of what you might think of as common sense and this can also be seen across regional, organisational and operational cultures.

I have also alluded to the principle that one can fly in the face of the common sense and still be right, so where does this leave us? I would argue that, in addition to raw instinct, the workplace audience will also need some form of technical competence. This will also facilitate the risk reduction required, particularly where only 'authorised' personnel are permitted to perform a task. Here I would commend another acronym and suggest that you ensure that all competent persons are (KEPT), as follows:

Knowledge – The theoretical or practical awareness or understanding of facts, information and skills acquired through experience or education. However, knowledge is not the same as wisdom, as the latter normally involves a healthy dose of perspective and the ability to make sound judgments rather than simply knowing.

In all walks of life, knowledge of both the task at hand and how to perform it safely is essential.

Experience – The performance or observation of tasks, activities and/or events, normally over time, which leave lasting and hopefully positive impressions.

The level of experience must be appropriate to the task(s) being undertaken. It is not just about being time-served, as sometimes lessons are learned through a single experience.

Personality – A characteristic way of thinking, feeling, and behaving, which embraces moods, attitudes and opinions. In relation to the personality required of a competent person, some positive personality traits would include (not an exhaustive list) the following:

- Adaptability and compatibility
- Compassion and understanding
- Drive and determination
- Patience and being conscientious
- Honesty and taking responsibility for your actions

- Wisdom and the courage to do what is right in tough situations

These behavioural characteristics, amongst others, distinguish one person from another and can be both inherent in an individual and/or acquired.

If you learn as much as you can, put your knowledge to the test by trying new things and then analysing and reflecting on your experiences, you will gain wisdom and become a wiser person.

Training – The acquisition of targeted knowledge and skills to fulfil specific requirements of a task, function or role. Skills training can also be used to re-educate and retrain people, whenever new technology, processes or systems are introduced or when someone is promoted or changes their function.

You should be aware of the wide variety of training methods, whether formal (accredited) or informal, face-to-face or remote (online), as follows:

- Lectures
- Group discussions and tutorials
- On the job training and toolbox talks
- Practical training exercises
- Tabletop exercises and role play
- Simulator training
- Coaching and mentoring

These methods should embrace each of the learning styles, being visual, auditory, reading and writing and kinaesthetic (hands-on), wherever possible.

In short, being competent is about having or being provided with the skills, knowledge and expertise needed to discharge the responsibilities of a particular role and that includes achieving a good standard of ethical behaviour. Without the necessary competence, in full or in part, does not mean that an individual is incompetent but simply that she/he might not 'yet' be competent. For instance, a new starter in an organisation or team is unlikely to start by being fully competent.

It may well be your task to provide them with the appropriate information, training, instruction and supervision to bring them up to the required level of competence throughout the duration of their responsibility. It is categorically not just a matter of having the appropriate qualifications.

Every parent or guardian, director, manager or worker has the moral obligation, at least, to recognise their own levels of competence, to address all the risks involved throughout their life, whether in the home, at play or in the workplace. You will need to use your wisdom and courage to apply the right control measures to manage those risks.

You can let someone else do it

Most of us carry insurance for our private vehicles to comply with the national legal requirements. In reality we are actually transferring some, though not all (ie excesses, restrictions and exclusions etc), of the financial liabilities that arise following a road traffic collision to someone else – your insurer.

In any insurance contract, the insured risks are transferred from you, or your organisation, to the insurer. In return for the insurer, or more accurately the underwriter, agreeing to pay out should your liability risk, or part of it, materialise they will charge a fee (premium). When it comes to things like private vehicle insurance, this premium is calculated by considering a function of several variables like age, type of employment, medical conditions etc.

Many other forms of insurance are also available, designed to cover the potential for loss or failure of some degree or another. In fact, you or your organisation can insure against virtually anything, from the effects of a fire or flood in your home, to your annual community or sporting event being rained off, to your organisation's failure to provide proper advice to a client. However, if you are not aware already, the ability to transfer liability risks is restricted to civil wrongs (torts) only, as it is impossible to insure against criminal wrongs (crimes).

Before accepting the risk, a professional person or company, known as an actuary, will have analysed the financial consequences of most known risks. They use mathematics, statistics and financial theory to study uncertain future events, especially those of concern to insurance and pension providers. The actuarial findings are then used to ascertain the correct premium that will be levied in the event of the risk occurring, and potentially more important, the failure of the controls to mitigate effect of the risk outcome.

> *"Yes, risk-taking is inherently failure-prone. Otherwise, it would be called 'sure-thing-taking'."* – Jim McMahon (Chicago Bears quarterback)

Another form of risk transfer is the delegation and acceptance of task(s) to another individual or organisation. Whenever you delegate tasks though, you must ensure that the recipient is made aware of the responsibilities and risks involved. It is very important for you to understand that you will never be able to delegate accountability for a task that is your responsibility to fulfil. If the person or organisation to whom you have delegated a task fails in its performance, it is still down to you to explain why it failed.

When you delegate you must always stay in the information loop. You must always make sure that the recipient of the transferred responsibility is fully competent to undertake the task(s). It may require additional training and support. Always remember that there is a real difference between delegation and abdication. If you simply dump it on them and walk away, you will be caught out when it goes wrong and called to answer to a higher authority. Heaven help you when asked, "What happened there?" and you have no clue. So do not just fly in, make a lot of noise, dump on someone and fly out again. That is what seagulls do. It makes an awful mess and just proves to everyone that you do not know what you are doing.

The same thing goes for subcontracting. This is where you or your organisation delegates, through a contract for service, to a third party, all or part of your own or your client's requirements. You are still accountable for the primary contract and will be expected to monitor the performance of that contract, as well as the standards agreed. It would do you no good whatsoever to simply say it had nothing to do with you when something failed or did not go according to plan.

In fact, the contract would probably preclude any such abdication and would hold you responsible, regardless of what your subcontractor did or did not do. This 'vicarious liability' is part of the strict common law doctrine of 'agency', where the actions or inactions of the subordinate are the responsibility of the superior. It would then be up to you to counterclaim against the subcontractor in the event of any damage or claim. That is why you must undertake robust contractor compliance checks, which includes making sure that appropriate and sufficient insurance cover exists should any failure occur that leads to harm. Harm, in this sense, is used in a very broad sense to include personal injury, property damage, loss of services, business interruption and reputational damage etc.

When you consider transferring a task, with its inherent risk(s), to any person or organisation, either inside or outside of your own organisation or team, always remember that you are only handing over responsibility for the task(s). Final accountability for the efficient, effective and safe performance of the task(s) will remain with you, right through to a satisfactory conclusion. Should it go wrong for lack of support by you, the recipient will simply need to point out that you failed to provide the necessary information and training, instruction and supervision. I have known circumstances when the recipient has simply said something like, "I told them I couldn't do it and I was just told to get on with it."

Do not be a seagull as no one likes to be dumped on or set up to fail.

Sometimes you will just have to accept it

Every single time you get out of bed and open your eyes, you are accepting risk. If you have young children, just going to the bathroom can be full of hazards, with toys lying around, especially that miniature dumper truck at the top of the stairs. Try transferring that risk to your four-year-old, or getting your partner to carry out a risk assessment before cuddling up for the night, let alone any thought of banning toys in the house. You have no choice but to grin, bear up and get on with life.

> *"Live as if you were to die tomorrow. Learn as if you were to live forever."* – Mahatma Gandhi

Sometimes, the inevitable is foreseeable and sometimes it is not, as with the so-called acts of God, although natural disasters are, generally, foreseeable. The timing is not, but the events repeat themselves through history and the global society is learning how to both forecast and deal with them more effectively. We still have a long way to go but nature will always win in the end.

Disaster planning and recovery is about learning some very hard lessons from past events and figuring out how to react, mitigate and recover. There is also the potential for turning adversity into advantage, by improving the lot of those who have suffered after the event has been dealt with. Although not the purpose or subject of this book, disaster planning is a very important consideration when it comes to dealing with the inevitable risk(s) to life on a large scale.

At the time of writing this book, we are all emerging from a pandemic. Although some say the catastrophe, of devastating proportions, was created by us humans, it is still interesting to note some of the common (world-wide) sense responses. Social distancing, strict personal hygiene and re-engaging with the community spirit all helped to restore social cohesion on a local, regional, national and global scale. Obviously, the vaccines, developed through global cooperation, ultimately allowed us to beat the virus, but getting through the first two years identified the very different versions of the common sense within each country.

Only you can decide whether you believe that the various country leaders listened to the common sense of the people, or just the scientists. It is worth returning to Sir Francis Galton again and his three caveats. First, each individual member of the crowd must have their own independent source of information, whereas there were so many voices, including social media. Second, they must make individual decisions and not be swayed by the decisions of those around them, whereas the scientists were driving the narrative. Third, there must be a mechanism in place that can collate their diverse opinions, and again, both the mainstream and social media expressed extreme bias in many directions.

My recollection though, from the outset, was that the common sense suggested that a total and immediate lockdown would have stopped the virus from spreading and all effort should have been made to protect the vulnerable. For me, it is hard to comprehend why, out of a total world population (2020) of 7,842,285,000, the loss of life at the end of 2020 (I know the final count was not yet in at that time) was 2,202,993 (or just 0.03%). Those who tested positive were 102,132,931 or 1.3% of the total world population, so the survival rate was extremely high, even before the vaccines became available in early 2021. I am not suggesting for one split second that we should have just accepted the death toll, but from these figures the lack of global cohesion and leadership meant that 99.97% were adversely affected by the effects of various lockdown regimes. I will leave you to make up your own mind on that one.

However, I will return to a more micro-level response to everyday activities. Basically, we have no real choice in any event, so we must continue to accept the inherent risks and opportunities that are going on all around us, every day of our lives. Getting up and getting dressed, cooking our meals, caring for our loved ones, travelling to and from our various destinations. This may come as a shock to you but I would even go as far as to say that you have no choice but to accept or tolerate many of the risks in the workplace.

We tolerate risk in different ways

Risk tolerance is your willingness, or the willingness of your family, friends and work colleagues, to take risks, which is based on several influencing factors. Understanding these factors is key to improving safety culture or, in the workplace, safety performance. I believe that there are three distinct sensory/cognitive processes that occur when assessing hazards and determining how much risk they present to us, as individuals, teams or even whole organisations, as follows:

- Sensory inputs received about acts and/or conditions that could lead to harm – *has the risk been identified (freeze)?*

- Sensory processes determine the nature and extent of the harm – *has the risk been understood (assess)?*

- Cognitive processes decide response – *has the risk been dealt with (fight or flight)?*

If you, your family, or your team decide to accept the risk, you will carry on. If you decide, sometimes in a split second, that you have been confronted with an 'at risk' situation, you must then determine whether it needs to be avoided, mitigated or transferred. The risk-assessment process can take hours or days, sometimes weeks, or it can be performed in a split second. When undertaken instinctively and quickly, it forms the basis for dynamic risk assessment (DRA), sometimes called last-minute assessment, and is how humankind has survived for thousands of years.

DRA has been frowned upon in the workplace until relatively recently, due to the potential lack of evidence to support the process. However, this has now become a very acceptable workplace methodology but only when it is formalised into a management system, alongside both the 'generic' and 'specific' risk-assessment processes. Without that formalisation, it is simply viewed as 'taking a punt' with your fingers crossed behind your back. So, before even considering formalising the adoption of this very natural DRA process, you will need to establish three Cs, as follows:

- **C**ommitment from the top – The senior management team will need to embrace the fact that traditional risk-assessment documentation will only be produced for the 'generic' (corporate or high level) and 'specific' (departmental or routine operational) risks. The 'dynamic' (worker level) risk-assessment documentation will not 'necessarily' be produced until after the activity. This will require trust in the workers, as this is predominantly a worker level function.

- **C**onsistent training – DRA training should be undertaken throughout the whole organisation, including those areas where traditional risk assessments are suitable and sufficient (administration and support services), in order that everyone understands the principles.

- **C**ommunications – The results of the DRA and performance of the relevant activities will need to embrace a feedback (debrief) loop and sharing of knowledge.

Your individual perception of risk is also an important factor. This is your ability to understand how an unsafe act or unsafe condition (hazard) could result in harm. It is, fundamentally, dependent on your background and competence, including the level of hazard recognition training you may have received. Your ability to predict the consequences of being exposed to the hazard(s) will also be determined by your level of tolerance (acceptance), which is generally influenced, adversely, by the following factors:

- Overestimating personal capabilities – Strength, agility, reaction time etc, together with an overreliance on the personal knowledge and experience of team members.

- Complacency – Overfamiliarity with a task or working on auto pilot and blind to potential hazards.

- Overconfidence (equipment) – Excessive or unwarranted trust in tools and equipment, especially in activities for which they were not designed.

- Overconfidence (PPE) – Higher levels of associated risk are accepted when the safety equipment provided for one set of circumstances is believed to prevent all forms of harm, under all circumstances.

- Underestimating seriousness of the consequences – Not considering the reality of the potential outcome(s).

- Historical experiences – Personal reality can cloud judgment, either by accepting higher or lower levels of risk, based on the outcome of a previous event.

- Peer influence – Risks accepted by peers, supervisors and mentors will directly impact the level of risk individuals will accept.

- Confirmation bias – Conviction or perception that the risk is under control and safe, despite the actual reality of the situation.

- Risk v reward – Personal reward balanced against the personal cost of participating in an 'at risk' activity. Risk tolerance will increase to dangerous levels where direct (financial gain or task and finish practices etc) and indirect (prestige or promotion etc) rewards are disproportionately high.

The key to dealing with all risks

The subject is vast, but in simple terms common sense dictates that a robust and practical mantra is required. You, your family, friends, work colleagues and the organisation for which you may work will need to actively implement a relevant process of risk management, as implementation is the single most important key to preventing harm.

> *"Action without vision is only passing time, vision without action is merely daydreaming but vision with action can change the world."*
> – Nelson Mandela

The reason why the courts have needed to get involved in this area of human failure, or lack of effective implementation, is because those who should have cared did not. Even before 'statutory' requirements were brought into force, civil courts had already determined, through 'precedent', several simple tenets relating to safe procedures. This case law has been developed over the years, where incidents of harm have led to litigation against those who failed to care for those who should have received the care. These cases have also led to various pieces of primary and secondary legislation around the world, which require all those responsible for the care of others to provide all the following:

Information – The collection of data, intelligence and material necessary to develop knowledge, which is normally obtained from investigation, study or instruction. In the context of health and safety it helps us to understand the potential hazards and risks, as well as the appropriate health and safety measures to minimise those risks. Remember though that:

"Data (good, bad or indifferent) gives rise to information (we are overloaded with information), which in turn gives us knowledge (dependent on the quality of the original data) and then we acquire wisdom, but although information and knowledge is in abundance, wisdom is in short supply." – Me (well, my twist on the DIKW pyramid used in information science, which was first mentioned in a play by T S Eliot in 1934, at least)

Training – The act of teaching, by a 'competent' person, a particular skill or behaviour, using a range of teaching methods (see also 'training' in relation to 'competence' above).

Instruction – A statement, command or sign (eg pictogram) that describes what to do or not do, or how to do something.

A safe operating procedure (SOP) or safety manual will contain a set of written instructions that identify the health and safety issues that may arise from plant (machinery, equipment and tools), place (site, building or location) or process (activity or procedure) and should identify, as a minimum:

1. Any hazards and risks involved.

2. Any inbuilt control measures.

3. Any specific training or qualification required to undertake the task.

4. The personal protective equipment (PPE) to be worn.

5. Any actions needed to reduce the risks.

Safe operating procedures do not replace the requirement for training but may be used to supplement or guide the training process.

Supervision – The act or function of overseeing something or somebody and although any person who performs supervision is a supervisor, she/he may not always have the formal title of supervisor. Effective supervision should also help monitor the effectiveness of the training received and whether workers have the necessary capacity and competence to perform their designated function(s).

Supervision is an integral part of controlling risks, although its nature and degree will vary according to the risks associated with the activity, the environment and the degree of competence of the worker being supervised. For instance, a newly appointed or young worker, who is not yet competent, will need close supervision. A competent worker may well only warrant remote supervision, although there should never be a situation where workers are devoid of any supervision, even if they are working away or out of sight.

It is important to remember that in most cases involving either the civil or criminal courts, it is the lack of supervision that normally determines if the appropriate level of care has been provided.

So, a very short precis of the core message of this chapter would be that: *"(IT IS) really important and 'competent' persons must be (KEPT)." –* Me

And as you can see, all these terms are intrinsically linked.

CHAPTER 8

SELF AND FAMILY FIRST

You must look after yourself before others

I am sure that you will have received the instruction from a member of the cabin crew when flying that in the event of a sudden decompression of the cabin, put on your own mask before that of your child, which sounds counterintuitive when you first hear it said. However, spending time putting on someone else's mask might incapacitate you through lack of oxygen and then you will be of no use to anyone.

A similar analogy can be used when it comes to evacuating premises where someone with a mobility issue, like being confined to a wheelchair, forms part of the queue. When asked, the common sense would suggest that as most people think emotionally, this person should be evacuated first. What do you think?

Well, logically, the presence of the slowest moving person at the front of the queue will hold up everyone behind, so it is now accepted practice that these people are taken out last of all. Again, this is exactly what happens when aircraft are being embarked and disembarked, with the less mobile getting on first and off last. This does not mean that they are left to their own devices.

In modern and/or recently refurbished buildings, a relatively recent requirement is for a disabled person's refuge to be provided. Its purpose is to facilitate that placement at the back of the queue but in a place of relative safety. This could be a dedicated space on each landing of a staircase or a fire-protected enclosure in the centre of a building with access either way to a fire-protected staircase. The principle involved is that a fire marshal or personal helper (buddy) stays with the vulnerable person and assists them out of the building once the main body of people have passed by, and they must not be left in the refuge to fend for themselves.

These flight mode scenarios (excuse the pun) have become accepted custom and practice and now the common sense approach to helping those less able than others. There are also various extremes of the fight mode, where you will also need to ensure that you are fit and healthy first. The obvious examples would be military or firefighting operations. In fact, there could be nothing worse, surely, than being rescued from a burning building by an unfit firefighter. It is hardly surprising, therefore, that firefighters spend a significant amount of their time undertaking physical training exercises. This is not just about honing their skills but also their bodies. Like the military, you will normally find that firefighters and other rescue services work hard and play hard.

In both cases, as well as all forms of warrior-type professions, like lifeboat crew and mountain or cave rescue, these individuals love the adrenaline buzz. I can say, from personal experience, that there is nothing better than running into a burning building when everyone inside is running (sorry, walking purposefully) out. Sometimes though, the 'red mist' can drop and cloud out sound judgment. In those circumstances, it is important to know when this is happening. Intensive training, in simulated conditions that reflect real-life scenarios, is essential. Therefore, military personnel take part in war games, and firefighters undertake simulated 'hot' fire training. Worst case scenarios are used to test soldiers, firefighters, lifeboat crews and rescuers to their limit. They must be able to look after themselves and members of their team so they can rescue those who need rescuing.

It does not always work out well though. Take the case of Lieutenant Colonel Herbert (H) Jones (chapter 4), the Falklands war hero, who some say 'saw red' when he led a suicidal charge against an Argentine machine gun emplacement. You might need to ask yourself was this blind rage or heroism above and beyond the call of duty. Only those who were there on that fateful day can truly answer that question but all we can say, for sure, is what was cited for his posthumous Victoria Cross. Amongst many other things it read, *"The display of courage by Colonel Jones had completely undermined [the enemy's] will to fight further"*, and this shows that his actions became one of the deciding factors in winning the Falklands War.

He led from the front, not always the best place for a commander, and died for his decision, but what if the enemy had not surrendered? How would the battle have turned out then? Would it have emboldened the enemy to fight harder?

I am in no way a Shakespearean scholar but in his play *Henry IV (Part I)*, Falstaff pretends to be dead on the battlefield but after Prince Hal leaves the stage, Falstaff rationalises that:

> *"The better part of valour is discretion; in the which better part, I have saved my life."*

> Or, as the saying goes, *"Discretion is the better part of valour."*

It was meant to have been taken as a joke. The irony though is that although truly courageous people may also be cautious, caution is not, in itself, the most important characteristic of courage. Perhaps it should be taken to mean that discretion is better than rash courage?

> *"Take calculated risks. That is quite different from being rash."*
> – General George Patton

The needs of individuals, teams and tasks

Although one of the UK's foremost authorities on leadership, John Adair (b.1934) also incorporated the idea that the development of the individual within a team setting would help achieve a designated task. He argued against the 'great man (or woman)' theory of leadership, where charisma, personal power and rhetoric were thought to be the recipe for success, as being inherently unreliable. I will return to the subject of leadership as a key component of safety performance in the next chapter.

Adair suggested that, in essence, if an individual's needs were not met, the team would suffer and performance of the task would be impaired. I do appreciate that the team, collectively, also has needs and generally, if these are not met then the individuals in the team will not be satisfied anyway. However, I am more interested in you at this stage, as without you there is no reason to continue.

Your family are your most important team though. I have always taken the view that no matter what, family must always come first. In addition to your loved ones being the strongest pull on your emotions, family issues can, for the same reason, get in the way of you concentrating on the task(s) at hand. This was driven home to me very early in my fire service career when receiving my first posting. I was told that it was brigade policy never to station anyone on their home ground, or anywhere near it, just in case I received a call while on duty to my own home. Sound advice, although I was also to find out, much later in my service career, that it was not a foolproof policy. Incidents involving family members did not always occur at home and sometimes fate plays some funny tricks.

As a senior fire officer based at Euston Fire Station, I was routinely on call over quite a wide area of London, covering the City of London and City of Westminster to the south and the boroughs of Barnet and Enfield to the north. I was coming to the end of one of my shifts and looking forward to seeing my extended family, who were arriving from Australia, later that evening. Having left Euston just as my shift ended, I received a non-urgent pager call which I ignored as I was now off duty. It kept repeating and as I was now out of my operational area as well, I kept ignoring but listened in for radio traffic, just in case. Eventually, I was called by radio, even though the control room staff knew that I was off duty. Anyway, when I responded, I informed the radio operator that I was both off duty and out of area. I was simply told to report to the nearest fire station and contact the control room by landline.

I duly pulled into Hammersmith fire station and, as is customary, summoned the officer in charge to request use of the phone. When I got through to control, to my horror I was informed that my wife, who was collecting the family from Heathrow airport, had been involved in a car fire. All I could ascertain from the control room officer was that fire crews were in attendance and a 'Stop' message had been received, indicating that no assistance was required. I was advised to proceed at normal road speed, as I was obviously not on the attendance and out of area in any event. All I could think about, with the scant information provided, was the safety of my wife and family. Nothing else mattered.

That 'red mist' descended and off I went, with 'blues and twos' up on to the motorway and out towards the airport. I switched channels and listened out for more information about the incident but there was none. It transpired that it was a small engine fire (my wife's car) that had been extinguished by the car park attendant, just as the fire engines arrived. Mobile phones were not in wide circulation back then and I was unaware of anything. I kept going at speed until I passed a gleaming white Rover car. It immediately pulled in behind me and stuck to me like glue. If it had got any closer, I think it could have syphoned petrol from my fuel tank. It was the police in an unmarked car and I was in trouble.

Luckily for me, I eventually heard some radio traffic that the fire engines were leaving the scene, and lo and behold, there they were crossing over the motorway on one of the bridges. I used that as a catalyst to pull over on to the hard shoulder and pretended to talk to my control room. Within a few seconds, one of the officers knocked on the front passenger window and beckoned to me to open the door. Luckily again, I was still in full uniform with all my regalia but that did not prevent the inevitable question, "And where would you be going in such a hurry then?", and after a very short pause, "Sir." I was sorely tempted to be just a little flippant but refrained. Instead, I explained to the officer that I was attending an incident, at the airport.

The trouble was that I had told a porky (pie = lie) and if I had been rumbled, it would have been a serious disciplinary matter. If it had gone that far it would have had the potential of my being dismissed. My being a serving magistrate at the time would not have been helpful to my cause either. However, once I started I felt that I had no choice but to continue. To get myself on to the front foot, I called the officer's bluff. I suggested that he contact his own control room to verify my story. The adrenaline was really flowing now. My luck was in that day though. Having pointed out the two fire engines earlier, he concluded that I must be telling the truth. He politely bid me farewell and to let it go. Now, had I really been on that attendance, I might well have told him to mind his own business, but being out of area and off duty. I decided that "discretion [was indeed] the better part of valou".

Sometimes, you must do what you must do. You can rest assured though that I was risk assessing every single action, dynamically, throughout the whole of that episode. It is amazing just how quickly and clearly you can think in times of stress. As the saying goes, "all's well that ends well", and on this occasion, it did end well. I not only got away with a potential speeding ticket but also it did not go any further and my career was still intact. More importantly, my wife and the family were still at the scene waiting for the breakdown van and the car was transported back to a local garage. The luggage went with the breakdown van and I took the family home.

It does go to show, for me, just how much of a pull is exerted by our loved ones. The pull of the heart strings is always much more powerful to most people than the pull of both the team and the task.

We all need to have fun

I do love children. I used to be one myself, albeit a little while ago now. Well, my nearest and dearest tell me regularly that I still am. A child that is. I mentioned previously (chapter 1) that children have much more fun when they are out of sight of adults, and in play they learn.

Have you ever watched children during playtime? Nearly all of them are laughing. The younger they are, the more fun they are having. None of them are risk assessing consciously. Yet, they all seem to go back home at the end of the school day unharmed and having learned something new. We can all take a leaf out of their book by doing the same; as we grow up, we seem to have less fun and do more risk assessing. What I am trying to say here is that you can get your safety messages across very effectively by playing games. I am not talking about practical jokes here, especially if you recall the incident with the chair (see chapter 3), as that is a form of bullying.

One of my clients recommended me to a ship builder in Cornwall. The safety manager at the shipyard was finding it difficult to enthuse the workforce about safety and the safety culture was very cavalier. It was also very macho and even the female workers seemed to perpetuate the negative approach to any efforts to improve safety performance. This had led to some serious incidents, including a fatality.

While undertaking some preliminary workplace safety assessments, I observed that the workforce was full of canteen banter and practical jokes. I suggested that we use the natural desire for fun to stage a work fayre. Initially, there was some resistance although it was more a fear of the unknown than resistance but the senior management team bought into the idea. I toured the shipyard with the safety manager and together we recruited some help and formed a small steering group. It was based on a representative cross-section of the workforce and we only selected those with a willingness and flair for some creative thinking.

The first game was the human roller coaster. This was designed and constructed by the carpenters to assimilate different ways that workers could fall from height, safely of course. It was no coincidence that this would be the first game to be developed, as this was the cause of the fatality and still high in the minds of the workforce. As it started to get built, the rest of the workforce became more and more intrigued, rumours started to spread and I could feel the buzz of excitement. It also became one of the most enjoyable of all the games in the fayre, with plenty of opportunity to get wet and a great deal of fun. The eventual feedback was very positive.

We then developed an archery game, which required the participants to acquire knowledge of hazardous chemicals. These were represented by different coloured balloons filled with either water or a totally innocuous substance. However, the budding scientists managed to get them to perform like the various real things. In one instance it left one of the players, who made the wrong choice, smelling awful as well. He learned a valuable lesson that day that he will probably not forget.

A set of oversized cups and dice were used in an enormous team game of snakes and ladders. It was all about risk taking and chance and this became extremely competitive. The workshops were rigged with machinery and electrical hazards and again, teams competed fiercely to identify the most.

Just to add to the fun, the team also arranged for some regular fairground attractions, totally unrelated to health and safety. The shipyard even brought in an ice cream van and a good supply of 'tiddly oggies', otherwise known as Cornish pasties. The only currency that could be exchanged on the day had to be obtained by winning events at the games, which made it even more fun and competitive. No one went without though.

No work was done that day and it was just like *It's a Knockout* or *Gladiator*. The only twist was that the games were all designed to facilitate a written test the following day. Every single member of staff, from the directors down, including the canteen staff, all achieved a British Safety Council certificate of achievement.

It all took weeks of planning, which was also fun but, my word, it was so worth it. Everyone was talking about it for months. It was just like watching children in the playground and it worked. The problem was that it was so successful that everyone wanted to do something the following year.

Sadly, as there were about 500 workers, it was seen by senior managers to be prohibitive on such a large scale. Obviously, it lost a whole day of shipyard time, not to mention the planning and construction work, and I dread to think of the financial cost. However, the original steering group became the safety committee. It is still functioning and regularly plans similar but much smaller scale events annually. They have also devised an incentive scheme with adrenaline prizes, like skydiving, power boating, track days, skiing trips etc, which are all highly valued.

The workforce is also encouraged to submit details of unsafe acts and conditions. They are also directly rewarded if they recommend an accepted solution which then features in the incentive scheme and annual safety awards.

The number of significant harm events was immediately reduced to very low levels and has stayed there ever since.

So, fun and safety do work very well together.

When things go wrong

Life can never be all fun though, as you and I know only too well, and there will be times when you find yourself in desperate circumstances.

Your role, at home, at work or at play, is one where you are likely to be called upon, at some stage, to make difficult decisions and choices. You will not always be able to do it alone and this is when you really need to talk to a 'critical' friend, or 'buddy'. Being responsible for the safety of others can, sometimes, take a heavy toll. It can be, and often is, a lonely place, where you just cannot do right for doing wrong. It can leave you feeling very isolated. You will need to offload to someone. Sometimes, you will need a shoulder to cry on or just someone with whom you can bounce your ideas around.

It might not be your partner or another member of the family or team that you can turn to, particularly if the issue concerns them. It might also be to do with you but you must find a way through the issue. Once the adrenaline of the fight or flight mode has subsided, it will leave you drained and when the issue continues, the stress levels change. No longer will you be getting an adrenaline buzz; instead, you may start to feel the stress that is brought on by worry.

When things go right, you can bet your bottom dollar that someone else will get the praise. You can also almost guarantee that you will be first in line to explain why someone has got hurt or something has been damaged. It goes with the territory. Just remember though, you cannot do everything.

All that is required is to get those for whom you are responsible to bed at night in the same condition as they woke up that morning. The art is getting them to perform their tasks safely, using their own tuning fork and not just walking by when they know, instinctively, that something is wrong. The same goes for you.

Your nearest and dearest will always be there for you, so just be open and honest with yourself.

CHAPTER 9

LEADERSHIP IS THE KEY

Even if you are not the leader you could find yourself leading

Let us go back to the situation when 'H' went down, in the Falklands. Someone else had to step in and that could, quite literally, have been anyone. In similar circumstances, not necessarily on the battlefield, it could be you.

If you are part of a team, the formal leader might just be on leave, or sick, or seconded elsewhere and you need to step up to the plate. It might even be that there is no formal leader in your small team but someone, potentially you, must take control, or the task(s) will not be completed.

How about your current leader? Is she/he effective or are you doing the leading for them and they just let you get on with it?

Whatever the circumstances and however it comes about, if you need to lead, then you must lead.

> *"Cometh the hour, cometh the man."*
> – Cliff Gladwin (Derbyshire bowler) 1948

> From an old English proverb – *"Opportunity maketh the man."*

> and earlier…

> *"The hour's come, and the man."*
> – Sir Walter Scott (Guy Mannering) 1815

It is not always easy

We have already spoken about the Adair theory of leadership (see chapter 8) and in very simple terms, that is a good starting point. Obviously, when the chips are down, it is the task that is the most important element. You will be judged, after the event, not just on whether it was performed but how it was performed. Did you get the task completed despite individuals within the team, or did you harness their individual and collective skills, knowledge and experience?

Did you lead from the front, or from the back and more importantly, would your team do it again?

If you recall, your skills, knowledge and experience are very important here but your competence as a leader includes your personal attributes, which is why your boss will want to make sure that you are (KEPT). With competence comes confidence and that is something that comes from self-belief. It is not about you believing that you are better than others. Rather, that by acknowledging your own faults and shortcomings, you can demonstrate a degree of humility. No one likes a smart-arse.

Even when I was the most senior fire officer at an incident, I did not always take on the leadership role. I was always accountable but if an officer of subordinate rank was in the throes of doing a good job and the incident was under control, I would allow them to run with it and just maintain a watching brief. On those occasions when I took over command, I would also listen to other officers and balance their advice, sometimes based on specialist expertise, with my own instincts. Every incident was different.

On one occasion, I attended a moderately large fire after receiving multiple calls to the incident where, on arrival, I observed the actions of a junior officer who was obviously out of his depth. I will not name either the officer concerned or the incident, for obvious reasons. He had already requested additional crews, even though the multiple calls triggered an increased attendance, which included me as a senior officer. As I arrived, I saw him running around like a headless chicken. It was a serious fire in a fish and chip shop, with run-down residential accommodation above and two families, at least, trapped inside. His inclination was simply to keep requesting more fire crews, even though the six he had requested had not yet arrived.

I managed to stop him in his tracks. I asked why he had not deployed one of his two crews to the rear of the premises to push the fire forward and away from the upper storeys. The other crew had not even been deployed to protect the single staircase and facilitate the evacuation of those trapped inside above the fire. Instead, he had deployed both crews to pitch ladders to the windows, without any covering hose lines. The projecting shop fronts meant that crews were working immediately above the fire and in jeopardy themselves.

I had already informed the control room that I was in attendance but decided not to formally take over the incident. However, after redeploying his crews, I reverted to him and suggested that he stand back alongside me and explain his actions as they occurred. The other fire crews then started to arrive and provided the additional resources to both complete the rescues and extinguish the fire. As soon as he calmed down, when events started to come under control, it was obvious to me that this officer, who I had not met before, was newly promoted and had never been confronted with this type of 'bread and butter' incident. All the training in the world can never take the place of real-life experience.

As soon as the incident was over, I returned to the fire station with the crews and joined in with the customary debrief. Now, I must tell you that a fire service debrief, like those in the military, is like no other. The first thing I did, after our cuppa, was to remove my rank markings and let the young junior officer (I wasn't that old myself) run the show. He showed his humility and immediately endeared himself to the other more experienced firefighters. Unfortunately, it did not stop the onslaught of criticism but he took it all on the chin and even gave some back. The debrief lasted for well over an hour and, from memory, closer to two hours, which was unusual.

I honestly believe it was one of the most constructive debriefs I had attended up to that point in my career, or since for that matter. No one took any prisoners but the learning was immense and the team spirit was positively enhanced. The young officer also showed a great deal of authenticity, by listening and being prepared to learn by his own mistakes. At the same time, he fed back his own leadership concerns. All those attending agreed that if they knew the leader was failing, they must provide the benefit of their professional knowledge and expertise, in a positive way. It is obviously very wrong for those receiving the orders in operational circumstances to question them negatively. It is only right and proper though that they should not allow someone in the leadership role to make unnecessary and, in this instance, potentially fatal mistakes.

It is also worth acknowledging that the chain of command starts on the first rung of the promotion ladder and goes right to the top. At every stage, while climbing that ladder, you will be looking for credibility. You will need to be caring of your team and trusted to keep your word, especially when the truth hurts. You must possess a high level of principles and be seen to adhere to them.

"A good head and a good heart are always a formidable combination."
– Nelson Mandela

What about self-deprecation; can you laugh at yourself?

People tend to laugh at those who are different. You should dare to be different, occasionally. Not all the time though, as those same people will probably not take kindly to a fool either. However, humour is a great leveller and good leaders know when humour is appropriate. More importantly, they also know when it is not.

Being a good comedian is all about timing, and the use of humour is no different. When things are getting a bit rough, you should consider performing a quick dynamic risk assessment. Revert to your cave and freeze just for a moment, to take a breath. Take two moments, in fact, while you take a few more deep breaths. The very worst thing you can to is to react too quickly and panic. Just remember that junior officer who, once he made his initial bad decision, found it got progressively more difficult to correct. Had I not been on the scene as quickly as I was, more decision making opportunities would have kept piling in on top of the first. The situation would have deteriorated rapidly and cumulatively. I know it is an old cliché but you must do your very best to keep calm and the 'take 2' rule should help you achieve that. The team is expecting you, as their leader, to keep calm. Nothing instils distrust, apprehension and even fear more than the vision of that 'headless chicken'.

The ability to forgive is also a leadership trait that is so often forgotten, along with the simple thing of saying thank you. Going hand in glove with praise where it is due is your tolerance of others. You had to learn, so you must allow members of your team to make mistakes as well, albeit not repeated ones. The debrief or feedback loop is the most powerful weapon at your disposal as a good leader. You will do well to use it often. If you can, get your team to practise the sharing of their failures on a regular basis.

> *"Failure is so important. We speak about success all the time. It is the ability to use failure that often leads to greater success. I have met people who don't want to try, for fear of failing."* – J K Rowling

Poor leadership is probably the most prolific cause of harm

Probably the worst trait of a poor leader, in my mind, is the reverse of forgiveness. It is the lack of trust that indicates that the leader is not prepared to empower their team to make the mistakes from which they will learn. So, if you are slow to praise and quick to criticise, you really should take that step back and look at yourself. I am sure that you do not like to receive criticism all the time and if you never get any praise your self-esteem will suffer. The same applies to your team, and if that attitude continues, everyone in the team suffers.

Just reflect, for a minute or two. Do you act like this as a parent or did your parents do it to you? Whether you or they did or did not, just try to forgive the flawed actions or omissions of others, especially your family and friends, just a little more. Believe me, you will soon see the positive results that will flow.

Next, in my book at least, is the lack of empathy that indicates that the leader just does not care about others in the team. As I stressed in chapter 5, care is not a 'duty'. Any leader that simply ticks boxes to provide evidence of compliance with that duty is a poor leader. Do you look out for members of your team like you would do for a family member or a friend?

Work colleagues are also a part of your wider family. If someone does not look well, do you enquire after their wellbeing like you would do for your children or parents? If they fail to check in with you when they should, do you start to worry?

If not, then maybe you would do well to revisit one of Charles Dickens' books, *A Christmas Carol*. This recounts the story of Ebenezer Scrooge, an elderly miser who was visited by some ghosts of his former business partner Jacob Marley. After their visits, Scrooge is transformed into a kinder, gentler (more empathetic) boss.

Unfortunately, history has shown us that poor leadership has been repeatedly and consistently identified as one of the primary root causes when major disasters are looked at under the microscope. In fact, relatively recently (2007) poor corporate leadership and governance following serious disasters in the UK led to a Corporate Manslaughter and Corporate Homicide Act. Now, an organisation is guilty of a criminal offence if the way in which its activities are managed or organised amounts to a gross breach of a relevant duty of care that causes a person's death. Individuals can still be charged with voluntary or involuntary manslaughter where the organisation is separately charged with corporate manslaughter.

It is obvious, therefore, that imposing a 'duty' just does not 'cut the mustard' in terms of incident prevention, even though it was intended to act as a deterrent. In all cases, the charges will come after the event and too late. Disastrously, for the whole of society, someone must die first. Whether the ultimate charge is brought about by negligence or recklessness, it is a state of mind and behind every organisation there are leaders. Are you one of those leaders?

Let us ignore the specific 'statutory duties' that are specified in the laws of the country where you live or work and return to the 'soft' leadership that determines how good or bad you are, at grass roots level. For instance:

- Are you considerate or inconsiderate?
- Do you focus on the strengths of others as opposed to their weaknesses?
- Do you promote the work of others in your team or take credit for it yourself?
- More to the point, are you secure in your own skin?

I believe that you must be a good leader or you would not have picked up this book in the first place, and would not have read this far. I am also pretty sure that you will have considered the traditional and obvious qualities of being a good safety leader already.

You can also become a truly great safety leader

You will now want to know what it takes to be a great safety leader. First and foremost, you will need to be an excellent communicator with a friendly demeanour. Then, there are a few more traits and actions that you will need to develop. You must always:

Make it personal. Be forthright about your commitment to safety.

Drive home the message, relentlessly. Make sure everyone knows it is a deep personal value and the issue that influences your decisions.

Identify what makes the organisation tick and demonstrate support. Establish and develop the mindset of both the workers and decision makers, to understand the safety culture and corporate safety values. Take priority action on the core safety issues.

Align resources with safety objectives. Ensure you have the people, financing and non-financial resources needed to succeed.

Align corporate and contractor safety objectives. Treat contractors in the same way as everyone else, as all life is sacred.

Align the workforce with their safety skillsets. Work collaboratively with your team and the workforce to get the most out of their performance by providing the necessary coaching, mentoring and training needed to succeed. Ensure you have extensive background knowledge of your industry and any specific safety projects.

Lead by example and do not tolerate unsafe behaviour. Make sure that safety is ingrained into your daily routine and that everyone knows that violation of safety instructions is subject to appropriate consequences. Always take a positive approach to your team, workplace and tasks and make the topics being discussed interesting and fun.

Drive proactive exposure reduction. Ensure that all unsafe acts and unsafe conditions are identified and measured so workers can focus on the prevention of underlying causes, not just injury prevention.

Expect everyone to care. Ask probing questions to understand how to get the workforce and contractors to improve the safety culture.

Actively seek feedback. Ask for constructive criticism and listen to what you are being told to make sure that your words and deeds match your own perception of what was intended. Extend team safety dialogue into the workforce, enjoy being around your team and always want what is best for them.

However, even if you are not yet undertaking these safety leadership functions, in my next (final) chapter, I will impart some of my own knowledge and ideas about what you can achieve. Hopefully, it will help you to fire up your inbuilt and intuitive personal attributes, which are possessed by all humans and which have helped humankind survive as a species for thousands of years.

CHAPTER 10

LISTEN TO YOUR BODY AND ACT

Messages we do not understand

Sometimes, we receive messages through our body that we just do not understand. whether they be premonitions, déjà vu or just that tuning fork that I have been mentioning. I am not going to go all spiritual on you, honestly, but I do believe that we all have a sixth sense. The most vivid recollection that I have concerns an incident where I very nearly came to grief in a serious fire. It was only the sixth sense of each member of the crew, at the same time, that saved my life.

As part of a specialist team of firefighters trained in the use of extended duration breathing apparatus, we were ordered to search a very large open warehouse. Several people were believed to be trapped inside. The whole area was completely smoke logged and other crews were deployed to find the seat of the fire. We were a team of six firefighters and were laying out what was called a guideline, from which we were able to search off in a perpendicular arc.

The conditions were awful. We could not see very much at all other than our next in line crew member. We could not get very much information from what we touched, as we were wearing proprietary (firecraft) gloves and heavy firefighting kit. We could not smell anything, as we were wearing full breathing apparatus. We did not want to taste anything, even if we could. We were all extremely hot and our core body temperatures were soaring.

The adrenaline was certainly flowing and that would have honed what was left of our five senses. All that was left, on the face of it, was our hearing and yes, we could just about hear our own mumbling communications. It was so very strange then that we all knew what was about to happen. Instinctively and at the same moment, we knew that the floor beneath our feet was about to give way. We instinctively moved closer and grabbed hold of the webbing on each other's harnesses. At that moment it went eerily quiet and we heard a very large crashing noise.

Luckily for all of us, the floor came away from the supporting structure on one side only. We were to find out later that it provided a ramp for us to climb up and get ourselves out. As the team leader though, I was at the end of the line of six and would have been the first to have fallen into the abyss. First, the team mate to my left, as there was no one to my right, reached out and grabbed my shoulder strap, then another hand reached out and I grabbed it in blind faith, while the other crew members hung on to them. The instinctive teamwork got us all out alive.

We were done and gave up the search at that point. All we could do was to follow the guideline back out of the building to safety. It was even hotter on the way out though, as the fire in the basement had now vented itself. We could see what a cruel ending fate would have dealt us had we not reacted to our tuning forks, not that we had any choice. It was pure instinct and an automatic human reaction.

You might well say that we all just reacted as we were trained. However, although training is a very important survival strategy, something else was at play here. The whole team agreed, even before we got to the debrief, that our tuning forks resonated in total harmony and that is why I am still here to write about it today.

Learn to understand

That incident was early in my career when I was a junior officer. I was to see many such events, including the fatal incidents that I mentioned at the start of the book.

However, I always knew there was something else going on around me but never paid it too much credence. If I was unable to see, touch or smell it, it was not real. Even after my second double parachute malfunction, I believed that I was indestructible and nothing could stop me. I was a firefighter, I jumped out of perfectly serviceable aircraft for fun and I rode a fast motorbike. I still do, although I have now stopped the firefighting and skydiving.

My stepfather on the other hand, being a member of the Zoroastrian faith, was very spiritual and he tried, in vain, for many years to extol the virtues of his deeply-held spiritual beliefs. It was only on the day that he died in a road traffic collision that he fulfilled his wish and I learned what he was getting at all this time.

His faith demanded that he be guided to his afterlife by priests who say prayers for three days and this started immediately on the day he died. I went to his house to sort out his effects while he was 'lying in state'. The house was empty as he had lived there alone since my mum died just two years earlier. I went into his bedroom and could not believe my eyes.

This happened at the end of November (28th) and it was bitterly cold outside so, obviously, his bedroom window was closed. The room was full of butterflies. Yes, that is correct, butterflies. I thought I was going mad or hallucinating at least. So, I tried in vain to get them out by opening the window and waving a pillow around but to no avail. I called out for my brother who was downstairs at the time. As soon as he ventured into the room, he was equally 'gobsmacked'. These butterflies were real and they would not budge. They stayed in my dad's bedroom for three days and three nights. We even showed them to other members of the family.

On the fourth day, Dad was cremated and after, when everyone came back to the house, we went to show off our butterflies. You guessed. There were none. Not even one. It dawned on me that it was my dad having the final word and my tuning fork went into overdrive. I was sure he was sending me a message and I started searching for answers to this nonsensical conundrum. I even had dreams about butterflies, although I have never seen one in late November since his tragic passing. I do still feel my dad's presence though when butterflies (especially a Red Admiral) flutter by and my tuning fork starts to vibrate. It transpired that in many cultures butterflies have a deep spiritual connection with the soul. Even in the Christian faith, the spiritual meaning resonates (that tuning fork again) with change, transformation and rebirth.

Obviously, this is my personal story and I do not expect you, or anyone else, to change your beliefs on the back of it. However, for me, my dad sent me a very personal and valuable message that day. We all have a sixth sense for a reason and we either use it or we lose it, at our peril. I always listen to my tuning fork whenever it vibrates, even softly. You may not believe in butterflies or messages from the other side but please, always listen to your tuning fork.

Confusion and noise

Have you heard the phrase 'she/he who pays the piper calls the tune'?

Well, if you think it has something to do with the Pied Piper of Hamlin, you are probably wrong. It most probably came from the use of bagpipes, an accompaniment for dancing, whereby the lord of the castle paid for the services of the piper and accordingly got to choose the tunes. The point is though, that in this modern world professionals sell their expertise, and depending on the client's wishes, within reason, will support that stance with their expert opinion.

Left to your own devices, you may well come up with a different opinion or interpretation, and strangely enough, that is what I am doing with this book. I have found throughout my working life that there are always three sides, at least, to every story (see chapter 2). To someone that has not travelled a similar path, it is possible that these different interpretations of the same set of so-called facts are more than just a little confusing. The more intellectual the argument(s), the more confusing. However, I have also been around long enough to realise that all these opinions and interpretations have common threads and it is those common threads that I have attempted to tease out. Hopefully, you will now have seen a more congruent set of basic principles that hold true for all and any application(s), anywhere in the world.

I do hope that you can use the most basic of human survival practices that I have discussed to cut through the information overload, the abundance of knowledge and myriad of opinions, so that you can be wise. Sometimes, you will need to fly in the face of common sense, so be brave. Try not to get distracted by the noise, which is all around us and detracts from the simple basic truths. One of the worst forms of distraction is, surely, the media and particularly social media.

How much store do you put by what you see and hear on the mainstream and/or social media?

The mainstream media (newspapers, radio and television) is particularly renowned for delivering bad news, as that is what sells. The media barons pay the piper and the journalists play the tune. Then there is social media which, although somewhat different, is still full of noise and false truths, like the conspiracy theories that abound. Everyone and their uncle will have an opinion and that is not confined to experts. Your task is to act wisely and even when the data is filtered, you must still ask yourself who is doing the filtering and for what purpose. Even the so-called 'fact-checkers' have been found to be playing the tune of a paymaster with vested interest.

We are all confronted, in all facets of our life, with information overload. Somehow, you will have to exercise control over the messages you are receiving and show wisdom in how this is interpreted. This will be even more important in your working life. Depending on your job title or role, and particularly in larger organisations, it is quite likely that functional team leaders or departmental managers will be offloading their raw data and/or unqualified information to you, especially if you are the health and safety manager.

In their minds, that is your function, to manage every little piece of information that has to do with health and safety. The word manager here though, is misleading and you would do well to consider changing it to something more akin to being an advisor. To manage is simply to ensure that something happens, but to many it is taken to mean that it is for you to cope. This is where you will need to explain, very simply and robustly, that you will teach them to cope for themselves but you will not do their caring for them. In fact, the more you do that, the worse the situation will become.

"Give a man a fish and you feed him for a day. Teach him how to fish and you feed him for a lifetime." – Lao Tzu (Chinese philosopher and founder of Taoism)

Every single team leader, supervisor, manager and director in the organisation is responsible for the care of themselves and others. It is the antithesis of your role to take that caring function (not duty) away from them, so just cut out that noise too.

Your task is to apply your knowledge wisely. When you apply your wisdom to the task in hand, you may well need to fly in the face of the common sense that prevails. Just listen to your tuning fork and allow it to guide you. Sometimes, you will make mistakes but that is all part of life's rich tapestry and even the best fishing experts have bad days.

Sometimes, your senses let you down

When your tuning fork forces you to freeze and the feedback you receive assaults your senses, it can be a bit like that red mist descending. It can be almost unbelievable and make you just want to switch off, or worse. When it happens, you will need every ounce of your inner strength to overcome the fear or rage.

I was very fortunate, in my early fire service career, by receiving a long-term temporary promotion at a very busy fire station in south London. The station officer was old school and not very far away from retirement. He was so well respected that he tended to do things that us young bucks would not dream of doing. On most night shifts, he took himself 'off the run' and left me at the age of just 22 years old to ride in charge.

On this one night, however, his tuning fork must have told him that something was about to go down. He rode in charge that night and I reverted to my role as second in command, riding in charge of the other fire engine. We had all settled down for the night and just after midnight the bells dropped. Out we went to an area of the station's ground notorious as bed-sit land, with many houses of multiple occupation and squatters. Not a pleasant place at all. On route to the call, control informed us that there had been multiple 999 calls and we built ourselves up for a 'working' job, as it was known, meaning that we had the potential for a serious fire and rescue operation. In addition, we all knew that the station officer rarely summoned assistance. He knew, anyway, that he had four fire crews to deploy and would invariably have it all under control by the time a senior officer arrived.

Indeed, as anticipated, a very serious fire was in progress and several people were trapped inside the building. My fire engine was the first to arrive and we pitched our main ladder across a line of parked cars to rescue several people who were screaming for help from a window on the second floor. The other crew got to work on the ground floor to protect us and the main staircase. Additional crews arrived from our neighbouring fire station and very quickly we got it under control. Then began the painstaking process of clearing up, but within a few minutes, one of the other crews found a baby, dead in its cot. As you will empathise, no doubt, this is probably the worst situation any firefighter can encounter. The morale of the crews took an instant nosedive.

Out of the corner of my eye, I saw what I later discovered was the father approach the station officer in what looked like a very aggressive manner. Although this would normally have been quite understandable in the circumstances, it still seemed unusual. For some reason, my tuning fork started to vibrate. I dropped what I was doing and went over to assist. What came out of the father's mouth assaulted my senses and those of the station officer. Now, it might well have been despair or traumatic stress following receipt of the information about his baby's fate. However, I was still horrified. When he asked the station officer if he could arrange for the recovery of a very expensive piece of audio equipment, I knew I had to act quickly.

The station officer was on the verge of swinging for the man and I just managed to grab hold of his forearm on the forward lunge. It took three more of us to hold him back. Maybe we should have just let him go for it, but we were all protecting the station officer, not the callous father. It is strange when you find yourself slightly removed from the centre of things, as any normal person would have wanted to take the man down themselves.

Sometimes, things just snap and your only hope in these circumstances is that your past wisdom has produced a supportive team. The task was done, the team performed well and the individual needed some very special support in his hour of need. Without it, we would have had to say goodbye to a very well-respected officer, who nearly did what any one of us would have done had we been walking in his shoes.

Believe in yourself

You are, obviously, the result of your upbringing, either through nature or nurture. More likely it will be a combination of both. We are all different but if you are aspiring to be a great safety leader you must have self-belief.

Having said that, you must also stand outside of your own ego. You must ask for and expect your closest work colleagues, as you would your partner in life, to say it straight. You may well be self-critical but you will just not see the whole picture. If you look in the mirror, you will see the physical you but you really do need to see and understand the psychological and emotional you as well.

Even now, my teams are only ever criticised by me when they fail to point out my failings. If they do not mention something that I should or should not have done, for fear that I would be upset, I will be upset. It is not really a matter of being upset though, rather a lost opportunity to improve myself. If you have not experienced it already, you will get a warm glow of satisfaction when you know that your team have got your back. When you have the confidence and self-belief to understand that you can also make mistakes, by seeking the help of the critical friends in your team, they too will experience that warm glow of knowing that you are a great leader.

There will also come times when you cannot do right for doing wrong, no matter how many people have your back. Sometimes, you and/or your organisation will be overcome with the inertia of the situation. You could even be told that 'elf and safety is getting in the way again; it is you that is the enemy and it is you that is in the firing line. Treat this situation like any other and revert to your tuning fork. Take a step back and reflect for two minutes. You will still have all your survival options open to you and recourse to all your powers.

> *"All power is within you; you can do anything and everything. Believe in that, do not believe that you are weak; do not believe that you are a half-crazy lunatic, as most of us do nowadays. You can do anything and everything, without even the guidance of anyone. Stand up and express the divinity within you."* – Swami Vivekananda

Three sets of voices will be screaming at this point. Firstly, your emotions will be in overdrive. That is your heart talking to you. Secondly, you will be trying to work out your next actions. That is your head talking to you. Thirdly, something will be vibrating inside. That is your tuning fork again, or your gut talking to you. All three could be right but generally you should always follow your gut. However, before you do, pass it by your head to work out the consequences and then pass it by your heart to feel the consequences.

This process does not have to be done in seconds and in most instances it will be best to sleep on it. You would also do well to pass it by your partner and/or trusted members of your team before doing anything rash. Remember, you are not doing this just for you and without you in the fight the team is potentially unprotected.

It can get scary

You will need to summon all your inner strength and courage to act for what you believe to be right and proper. The hard things in life are worth fighting for and you are now both competent and confident. I know, it does not always feel like that but you know you need to fight the end game.

One of my business bibles is *The Prince* by Niccolò Machiavelli. He was an Italian diplomat in the late 15th and early 16th century. This book, amongst others including the *Art of War*, brought him a reputation as an atheist and an immoral cynic, but others, like me, believe that he was the father of modern political philosophy and political science. He typified the end game strategy by suggesting that all people, including princes, were in pursuit of what was good for them and, accordingly, you just need to find out what 'floats their boat(s)'.

He suggested that men (women as well) intrinsically mistrust what they have not experienced themselves and the prince needed to either seek help or use force (preferably both) to achieve his innovations. I am not for one minute suggesting that you use force in the literal meaning of the word but he was talking figuratively. What he was alluding to was the suggestion that you must change people's behaviour first and their attitude(s) will follow, similar to forcing drivers to wear seat belts. Initially, the public safety message was treated with contempt, but after many years of enforcement by the police, very few would even think of driving off without buckling up.

"Grab them by the balls and their hearts and minds will follow."
– General George Patton (1944)

Accordingly, I would recommend *The Prince* as a worthy addition to your bookshelf. My other recommendation for you is my other business bible, also surprisingly called *The Art of War* by Sun Tzu. In the depths of despair, both of these books will equip you with some essential advice on strategic thought processes, as well as some battlefield tactics that contain some transferrable and less dangerous skills.

"If in the midst of difficulties we are always ready to seize an advantage, we can extricate ourselves from misfortune." – Sun Tzu (544 to 496 BC)

Similarly:

"To know in war how to recognise an opportunity and seize it is better than anything else." – Niccolò Machiavelli (1521)

A little bit more up to date and the general philosophy becomes:

"When fate hands you a lemon, make lemonade." – Dale Carnegie (1948)

Most people will instinctively avoid struggle because it feels like failure and that will scare them. However, the phrase "no pain, no gain" also holds true, especially for you, so you must simply turn adversity to your advantage.

Share the pain

Put your thoughts into perspective and just consider this. If you are freaking out about something, life is moving on without you and you will not achieve what you believe was worthwhile. What is the worst that could happen?

Your loved ones will still love you. Your true friends will still be there for you. Your work colleagues will still trust you. Life will continue and you cannot be flogged any more.

If you procrastinate, you will have just delayed the inevitable. I am not talking about laziness here, as procrastination is an active rather than a passive process. You will choose to do something else instead of the task that you know you should be doing, especially when it is an unpleasant one.

I know I said sleep on it but please do not take that to imply apathy or inactivity. In fact, if you are the sort of person that I think you are, you will probably refuse to listen to your tuning fork as well. If it continues, it can make you feel guilty and/or ashamed. It can lead to reduced productivity and if left too long can cause you to become demotivated and disillusioned with your work. In extreme cases, it can lead to depression and even job loss.

You have unfettered access to your loved ones, friends and trusted colleagues, so maybe just make a start by writing down your pain. If you just think it through in your head it will become a jumble, whereas you will need to be objective.

Remember when I said you needed to trust others to make mistakes?

Well, the same goes for you. Even if you realise that what you have done or not done is the cause of the problem, just forgive yourself and move on. Even then, it is worth sharing. You will not look stupid, you will look human. You might even find some ongoing practical support.

It might be worth going to someone in a more senior position and, no, that does not mean that you need to be a sycophant. The important point is that you must be sure that the person will give you both a frank opinion and the moral support needed to take your issue(s) forward.

Just do it

Nike's slogan – "Just do it" – instils a passionate sense of motivation and the avoidance of hesitation and regret. It will also help you to build inspired emotions in your team and might even help you.

So, a very simple final question. Do you want to make a difference?

If so, as I believe you do, you will need to make an impact. You will need to engender the passionate belief that you and your team will never ignore any unsafe acts or any unsafe conditions. It will also require that you proactively contribute, even if only in a small way, to a better future, which embraces efficiency and effectiveness in a safe environment. You must continuously promote, develop and support initiatives that offer solutions to the problems you identify or are presented to you. You must do this now by:

- Dedicating yourself to your purpose
- Committing to continually bettering yourself
- Engaging with as many influencers as possible, for the mutual benefit of others
- Investing your time and energy in what you can reasonably achieve
- Embracing critique and welcoming scrutiny
- Sharing what you know and even what you feel
- Helping others to grow and succeed

"A man would do nothing if he waited until he could do it so well that no one would find fault with what he has done." – Cardinal (Saint) John Newman (1801 to 1890)

So, forget the goal and just enjoy the journey but use your influence well.

ABOUT THE AUTHOR

Malcolm Tullett has an insider's view of risk and safety. Having served as a senior operational fire officer in the London Fire Brigade, he has seen the fatalities first-hand and played a pivotal role in clearing up the mess when things go wrong. Now, as a business owner, advising all types of organisation from small to blue-chip on health and safety matters, he understands the challenges of protecting the wellbeing and safety of staff and property.

Health and safety is a highly regulated and sometimes risky business! The introduction of the Health and Safety at Work Act in 1974 has helped drive down the number of workplace accidents but has not eradicated them. This book takes safety to a new level and explores the belief that work environments will only become less risky when individuals incorporate best practice with managed intuition.

Malcolm's interest and understanding of risk and safety started at naval college. He continued to learn and analyse the rationale and effectiveness of long-standing practices whilst studying for an HND in Nautical Studies. He joined his first ship (part of the Cunard fleet) in Durban South Africa at the age of 16 as a cadet navigating officer. Life at sea was replaced by life in a fire station and Malcolm's potential was quickly recognised whilst training as a firefighter in Southwark. Malcolm rapidly rose through the ranks and at 22 was at that time the youngest UK station officer with responsibility for 18 crew and safeguarding thousands of Clapham residents. After a distinguished fire service career, and being promoted to Assistant Divisional Officer, Malcolm became a senior safety trainer before starting his own successful health and safety business.

This book is a game changer and an absolute must for business owners, anyone working in the industry, or those who have responsibility for people and property.